Introduction to the Biology of Marine Life

Introduction to the Biology of Marine Life

Frank Parker

STATES
ACADEMIC PRESS
www.statesacademicpress.com

States Academic Press,
109 South 5th Street,
Brooklyn, NY 11249, USA

Visit us on the World Wide Web at:
www.statesacademicpress.com

ISBN: 978-1-63989-313-3 (Hardback)

Cataloging-in-Publication Data

Introduction to the biology of marine life / Frank Parker.
p. cm.
Includes bibliographical references and index.
ISBN 978-1-63989-313-3
1. Marine biology. 2. Aquatic biology. 3. Marine organisms.
4. Marine sciences. I. Parker, Frank.
QH91 .I58 2022
574.92--dc23

Table of Contents

Preface

Biology is a natural science which focuses on life and living organisms. It attempts to study their physical structure, physiological mechanisms, development and evolution. The marine life or organisms living in the seas and oceans are studied under marine biology. Some of the common habitats of marine life are coral reefs, seagrass meadows, kelp forests, tidepools, deep sea trenches and surroundings of seamounts and thermal vents. Marine ecosystems exhibit an inverted energy pyramid structure. Microorganisms such as phytoplankton, zooplankton, ctenophores, molluscs, arthropods, etc. are at the bottom of the pyramid while large organisms such as whales and sharks are at the top. Marine life is studied under various sub branches of biology such as phycology, ichthyology and invertebrate zoology. The topics included in this book on biology of marine life are of utmost significance and bound to provide incredible insights to readers. It elucidates new techniques and their applications in a multidisciplinary approach. This book aims to serve as a source of information for students and experts alike and contribute to the growth of the discipline.

To facilitate a deeper understanding of the contents of this book a short introduction of every chapter is written below:

Chapter 1- Marine biology is the scientific study of marine organisms, their behaviors and interactions with the environment. It also includes the study of the marine ecosystems and human interactions with these environments and species. This is an introductory chapter which will introduce the reader to all the significant aspects of marine biology.

Chapter 2- Marine vertebrates are the vertebrates or animals with a backbone that inhabit the marine ecosystem and environment. Marine or saltwater fishes, marine reptiles, marine mammals, seabirds and marine amphibians are classified under marine vertebrates. Each topic has been thoroughly examined and explained in this chapter to provide the reader with a better understanding of marine vertebrates.

Chapter 3- Marine invertebrates are animals that neither possess nor develop a vertebral column, and live in marine habitats. It includes all types of marine sponges, cnidarians, worms, echinoderms, molluscs and arthropods. This chapter closely examines these marine invertebrates to provide an extensive understanding of the subject matter.

Chapter 4- Marine habitats are the ecological or environmental areas that support marine life. Marine habitats form an important part of marine ecosystems as they are rich in biodiversity and support a wide variety of life forms. Marine habitats can be further divided into coastal and open sea habitats. This chapter delves into the realm of marine habitats and provides an in-depth explanation to readers.

Chapter 5- Corals are invertebrate animals belonging to the phylum Cnidaria and can be classified as either hard or soft corals. Hard corals absorb the calcium from seawater to form hardened carbonate structures which have developed into large structures called coral reefs, providing home to a large number of species. This chapter provides a detailed and close-up analysis of corals and the reefs they have help build.

Chapter 6- Marine microbes are tiny, single celled organisms that live in the oceans and make up a majority of the ocean biomass. They consist of all sorts of bacteria, viruses and fungi. Marine microbiology is the study of these organisms, their biodiversity, ecology and biogeochemistry. This chapter has been carefully written to provide an easy understanding of marine microbes.

Finally, I would like to thank the entire team involved in the inception of this book for their valuable time and contribution. This book would not have been possible without their efforts. I would also like to thank my friends and family for their constant support.

Frank Parker

Chapter 1

Marine Biology: An Introduction

Marine biology is the scientific study of marine organisms, their behaviors and interactions with the environment. It also includes the study of the marine ecosystems and human interactions with these environments and species. This is an introductory chapter which will introduce the reader to all the significant aspects of marine biology.

Marine Biology is the study of marine or aquatic animals, about their activities, life cycle, their interactions with the environment, and different facts related to the marine aquatic ecosystem along with its habitats.

Charles Darwin is best known for his theory of evolution, and James Cook is known as the father of modern marine biology. During the 18th Century, he performed several significant researches to advance the study of marine biology, which also encouraged other scientists in their inventions.

Marine biology is the scientific study of the marine ecosystem, including different types of oceans, tidal zones, saltwater swamps, mangroves, rivers, streams, etc. along their features, depth of the ocean beds, different types of marine life, which mainly involves a vast number of species, from the microscopic phytoplankton and zooplankton to massive organisms like aquatic plants, different types of fish, whales, and sharks that lives, survives, and exists, under the surface of the ocean. Overall, Marine biology also includes inter-relationship between marine life, ecology, and the environment.

The study of Marine biology shows a wide variety of adaptations which involves the physiological, structural, behavioral adaptations and life cycle of marine habitats. Marine habitats are older than the terrestrial habitats, as life originated in the oceans was about 3.1 billion to 3.4 billion years ago.

Marine Organisms and Adaptations

Adaptation is an evolutionary process whereby an organism becomes increasingly well suited to living in a particular habitat. It is not a quick process! Natural selection over many generations results in helpful traits becoming more common in a population. This occurs because individuals with these traits are better adapted to the environment and therefore more likely to survive and breed.

Adaptation is also a common term to describe these helpful or adaptive traits. In other words, an adaptation is a feature of an organism that enables it to live in a particular habitat.

Figure: Bottlenose dolphins: Dolphins are adapted to living in water. They have a streamlined shape and fins instead of legs. They also have blowholes on the tops of their heads.

Different Types of Adaptations

Marine organisms have adapted to the great diversity of habitats and distinctive environmental conditions in the marine environment. Adaptations are many and varied but they are generally grouped into 3 main categories: structural, physiological and behavioural.

Structural Adaptations

Structural (or morphological) adaptations are the physical features of the organism. These include things you can see, like its shape or body covering, as well as its internal organisation. Following are a few of the ways that marine organisms have adapted their physical features to suit a particular habitat.

Seawater is much denser than air – as a result, there are vast numbers of microscopic organisms suspended in it. Cockles, as well as many other bivalves, are filter feeders. They have adapted specialised siphon structures to filter these organisms and any other particles of food from the surrounding water.

Estuaries have quite variable conditions – tides, waves and salinity fluctuations affect the animals and plants that live there on a daily basis. Many animals, such as cockles, are adapted to live in these conditions. They have strong shells that protect them from wave action, drying out and the prying beaks of predators.

Coastal plants need special adaptations to survive. For example, many types of seaweed attach firmly to rocks so they are not swept away by waves. Their leaf-like fronds are tough and leathery, which helps protect them from being torn by the waves or dried out by the sun.

Seaweed: Seaweed is a type of algae and is normally grouped by color – green, red or brown. Seaweed is a primary producer and an important source of nutrients and energy in the marine food web. Seaweed generally grows in clusters called 'beds', and these provide an important source of shelter for many marine species, including mussels.

Dolphins are mammals, but they look very different to mammals that live on land, as they are adapted to living in water. They have a streamlined shape and fins instead of legs. They also have blowholes on the tops of their heads. They use these to breathe, rather than through their mouths and noses.

Physiological Adaptations

Physiological adaptations relate to how the organism's metabolism works. These adaptations enable the organism to regulate their bodily functions, such as breathing and temperature, and perform special functions like excreting chemicals as a defence mechanism.

Some marine mammals, such as whales, migrate over large distances and may spend time in a combination of arctic, tropical and temperate waters. To cope with these temperature changes, they are endothermic or 'warm blooded'. This means that they are able to maintain a constant body temperature that is not dependent on the surrounding water.

Slow-moving species have adaptations that help protect them from predators. For example, many marine organisms can only move slowly or not all. This means they cannot easily get away from mobile predators, and they have other adaptations to protect them from being eaten. These can include chemical defences in their skin, for example, sea stars.

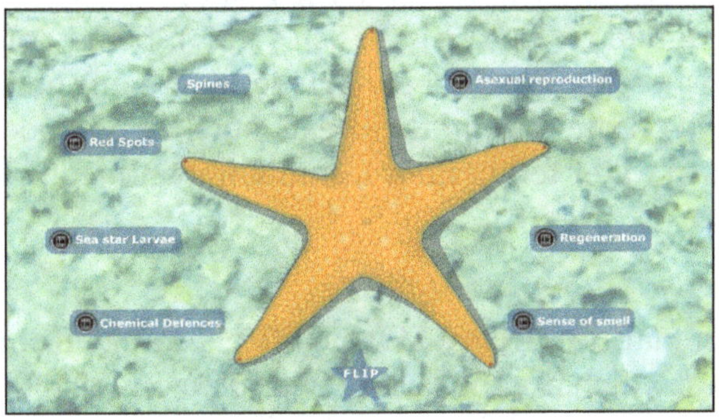

Figure: Sea star adaptations.

Sea stars have many weird and wonderful adaptations - including some unusual internal systems.

Behavioural Adaptations

Behavioural adaptations are learned or inherited behaviours that help organisms to survive, for example, the sounds made by whales allow them to communicate, navigate and hunt prey. Crab larvae use sounds to help them find suitable habitats so they can settle and metamorphose (change into an adult form).

Figure: Bryozoan colony.

Bryozoans form colonies that often look a lot like a plant. However, they are made up of hundreds of individual animals called zooids. Groups of zooids are adapted to play different roles in the colony.

Bryozoan colonies are found in high numbers on the continental shelf in New Zealand. They look like plants but are actually made up of hundreds of tiny individual animals that have banded together in order to more successfully find food and survive predation.

Marine Life

The phrase 'marine life' refers to organisms that live in salt water. These can include a diverse array of plants, animals and microbes (tiny organisms) such as bacteria and archaea.

Marine Life are Adapted to Life in Saltwater

From the perspective of a land animal like us, the ocean can be a harsh environment. However, marine lives are adapted to live in the ocean. Characteristics that help marine life thrive in a saltwater environment include the ability to regulate their salt intake or deal with large quantities of salt water, adaptations to obtain oxygen (e.g., a fish's gills), being able to withstand high water pressures, living in a place where they can get enough light, or being able to adjust to a lack of light. Animals and plants that live on the edge of the ocean, such as tide pool animals and plants, also need to deal with extremes in water temperature, sunlight, wind and waves.

Types of Marine Life

There is a huge diversity in marine species. Marine life can range from tiny, single-celled organisms to gigantic blue whales, which are the largest creatures on Earth. Below is a list of the major phyla, or taxonomic groups, of marine life.

Major Marine Phyla

The classification of marine organisms is always in flux. As scientists discover new species, about the genetic makeup of organisms, and study museum specimens, they debate how organisms should be grouped.

Marine Animal Phyla

Some of the most well-known marine phyla are listed below:

- Annelida - This phylum contains segmented worms. An example of a segmented marine worm is the Christmas tree worm.
- Arthropoda - Arthropods have a segmented body, jointed legs and a hard exoskeleton for protection. This group includes lobsters and crabs.
- Chordata - Humans are in this phylum, which also includes marine mammals (cetaceans, pinnipeds, sirenians, sea otters, polar bears), fish, tunicates, seabirds and reptiles.

- Cnidaria - This is a diverse phylum of animals, many of whom have stinging structures called nematocysts. Animals in this phylum include corals, jellyfish, sea anemones, sea pens and hydras.
- Ctenophora - These are jelly-like animals, such as comb jellies, but they don't have stinging cells.
- Echinodermata - It includes such beautiful animals as sea stars, brittle stars, basket stars, sand dollars and sea urchins.
- Mollusca - This phylum includes snails, sea slugs, octopuses, squids, and bivalves such as clams, mussels and oysters.
- Porifera - This phylum includes sponges, which are living animals. They can be very colorful and come in a diverse array of shapes and sizes.

Marine Plant Phyla

There are also several phyla of marine plants. These include the Chlorophyta, or green algae, and the Rhodophyta, or red algae.

Marine Ecology

Marine Ecology is the scientific study of marine-life habitat, populations, and interactions among organisms and the surrounding environment including their abiotic (non-living physical and chemical factors that affect the ability of organisms to survive and reproduce) and biotic factors (living things or the materials that directly or indirectly affect an organism in its environment).

Figure: Manado Tua volcanic island where the first Indonesian Coelacanth was discovered in 1997.

Marine Ecology is the scientific study of marine-life habitat, populations, and interactions among organisms and the surrounding environment including their abiotic (non-living physical and chemical factors that affect the ability of organisms to survive and reproduce) and biotic factors (living things or the materials that directly or indirectly affect an organism in its environment).

Marine ecology is a subset of the study of marine biology and includes observations at the biochemical, cellular, individual, and community levels as well as the study of marine ecosystems and the biosphere.

The study of marine ecology also includes the influence of geology, geography, meteorology, pedology, chemistry, and physics on marine environments. The impact of human activity such as medical research, development, agriculture, fisheries, and forestry is also studied under marine ecology. In some ways, marine ecology is more complex than the relatively straightforward study of a particular organism or environment because of the numerous interconnections, symbiotic relationships, and influence of many factors on a particular environment.

To understand the difference between marine biology and marine ecology, it may be useful to look at a community of organisms. A marine biologist may focus on behavioral relationships between the organisms in one particular species while someone studying ecology would study how the behavior of one organism influences another. An ecologist would also look at abiotic factors and how they influence that organism. A scientist studying community ecology might study a group of organisms to see how they influence other species and abiotic factors. The major subcategories of ecology are:

- Physiological ecology: The study of how biotic and abiotic factors act on the physiological characteristics of an organism and how the organism adapts to the abiotic and biotic environment.
- Behavioral ecology: A subcategory of ecology that studies which ecological and evolutionary dynamics is responsible for the way in which organisms adapt to their environment.
- Population ecology: The study of populations of organisms in a particular species and how the populations interact with their environment.
- Community ecology: The study of how species react to each other in a community.
- Landscape ecology: The study of how organisms interact with a particular landscape.
- Ecosystem ecology: The study of how energy and matter flow through ecosystems.
- Global ecology: The study of how energy and matter interact in the entire web of life on Earth.

The study of ecology in general includes all of the subcategories listed above as they apply to marine ecology, animal ecology, plant ecology, insect ecology, arctic ecology, tropical ecology and desert ecology.

Requirements for Sustaining Life on Earth

Earth has been divided by ecologists into four areas: the hydrosphere, the lithosphere, the atmosphere, and the biosphere. The hydrosphere refers to water on the planet, the lithosphere consists of soil and rocks, the atmosphere is the air, and the biosphere refers to all of the life on Earth. The biosphere can be visualized as a thin surface layer on the Earth from 11,000 m below sea level to 15,000 m above sea level, even though there are no permanent residents living in the atmosphere.

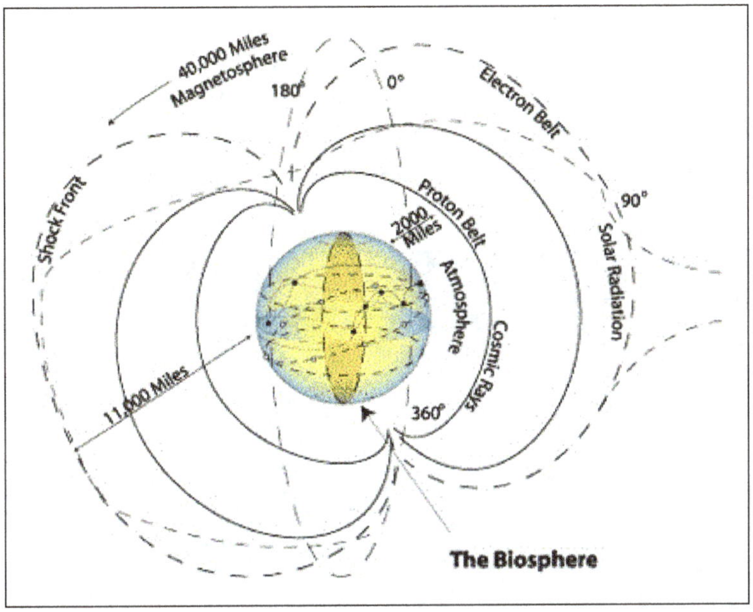

The first life on Earth was formed in the photic zone of the hydrosphere when organisms with more than one cell evolved in the deep ocean benthic zones. After the ozone layer formed, which protects land organisms from harmful UV rays, life began to evolve on land. After the continents separated and reformed, biodiversity began to increase as life began to adapt to new environments. Biodiversity can be observed at the genetic level, the species level, the population level, and the ecological level.

Abiotic elements like carbon, nitrogen and oxygen are present in great quantities in the biosphere. Phosphorus, calcium, and potassium are also present in smaller amounts. All are elements critical to the existence of life. Every element in the ecosystem transforms from mineral to organic forms and back to minerals and is never destroyed. Life depends on energy from the sun and the organisms that are capable of transforming light into chemical energy form the basis for the food chain. The process of photosynthesis converts light into chemical energy, resulting in the production of glucose and oxygen. Other organisms depend on glucose produced by photosynthesis for energy to fuel biological processes such as cellular respiration. During cellular respiration, organisms split glucose back into water and carbon dioxide. The breaking down of glucose releases and utilizes energy from the sun stored by photosynthesizing plants. The oxygen level of the Earth's atmosphere is largely reliant on the amount of photosynthetic activity and respiration going on in the biosphere. A buildup of elements in areas with a lot of organisms is prevented by circulation of the atmosphere with global air currents.

The Earth is full of cycles simultaneously occurring and interlaced. Water is cycled through the hydrosphere, lithosphere, atmosphere, and biosphere in relatively predictable movements.

The flow of basic elements and the stability of the Earth's climate and temperature depend on oceanic currents and the vast amount of water stored in the ocean. Ecologists often depend on computer modeling to determine how human activity can influence the intricate cycling of the biosphere.

Every organism is influenced somehow by every part of its environment. An ecosystem is any place or time where organisms interact with their environment. Ecosystems can be divided into the life in the area and the area in which life exists. Biocoenosis is the life in the area and the biotope is the environment by which life is influenced. One form of interaction in the ecosystem is the food chain, a system moving energy and matter through organisms and the environment. A micro-ecosystem is a small system like a fish with parasites. A mesoecosystem could refer to the coral reef in which the fish lives and a macro-ecosystem might be the area in which the coral reefs are located and how they interact.

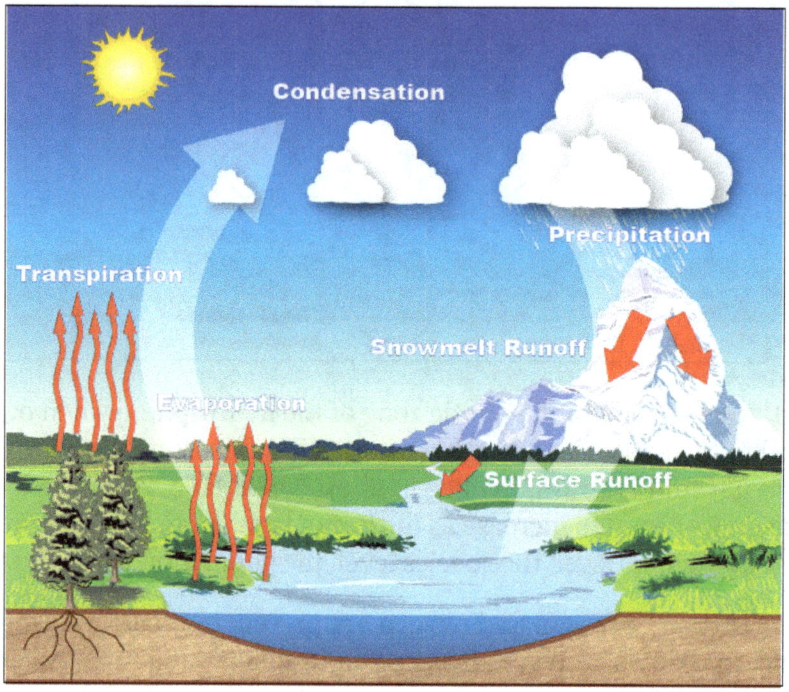

Abiotic factors include geographical, climatological and geological influences. Biotopes are determined by certain abiotic factors. Examples of abiotic factors include, water, air, soil, pH, salinity, temperature, amount of light, and even natural catastrophic events.

A grouping of populations of plant, animal, or microscopic organisms is referred to as a biocenose. When the number of organisms in a species becomes too small, inbreeding reduces the genetic diversity causing weakness in the species and possible extinction. The stability of a biocenose is also connected to biotic ecological interactions by organisms of the same species or organisms of different species.

Interactions between organisms of the same species are cooperation, competition, territorial divisions, and organization in the population. These factors are collectively referred to as intraspecific relations. The interaction of organisms of different species is referred to as interspecific relations and can include symbiotic interactions, competition, parasitism, and infectious disease. When two organisms occupy the same ecological niche, competition for resources can occur. It is important to determine whether the interaction is positive for species, negative for both, or positive for only one species. Another example of an interspecific relation that is negative for one species and positive for the other is predation. Predatory activities form the basis for all food chains. While predation is natural and provides nourishment for the predatory species, in some cases it can upset the balance of the food chain when the prey species is already overexploited.

Biogeochemical cycles are present when minerals and organic materials are used by organisms and sent out as waste. Ecosystems can remain relatively stable when untouched by catastrophic events, detrimental human activities, or other unusual occurrences. Homeostasis, or self-regulation of ecosystems, occurs when supported by natural control mechanisms.

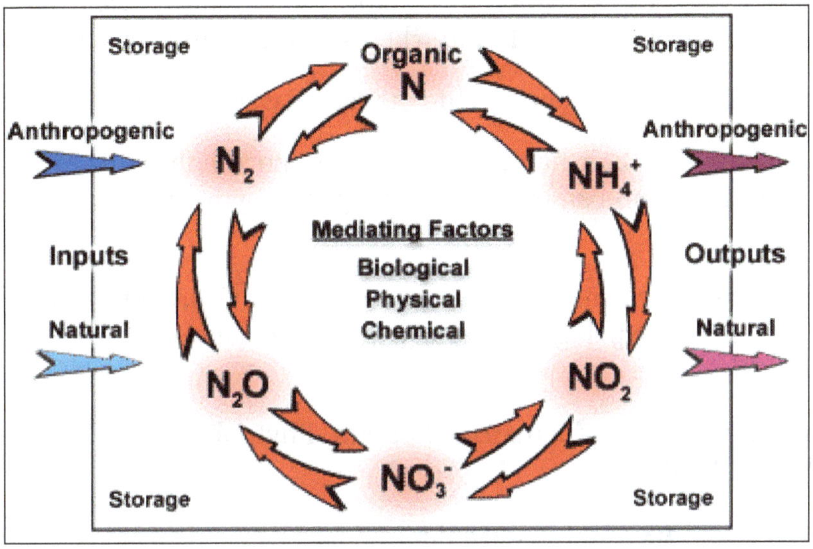

Figure: Nitrogen Cycles Project: Illinois State Water Survey.

A biome is an ecological area separated from other areas by certain definitive characteristics and relies on the interdependent nature of ecosystems. Water, among other elements, can cycle from one ecosystem to another. Organisms like salmon and freshwater eels often move from one ecosystem to another. The biosphere includes all of the Earth's biomes. An example of a biome is the photic portion of the ocean where sunlight is present and photosynthetic algae can be found. Biomes are divided into ecozones corresponding loosely with the continents and are further divided into ecoregions.

Species are related to each other through the role they play in the food chain as producers, consumers, and decomposers. Producers are photosynthesizing plants, consumers are herbivorous or carnivorous animals, and decomposers are organisms (such as bacteria) that break down organic material into minerals, which are eventually used by producers. There are more producers than consumers. The total amount of living matter in any place is called the biomass. When the biomass of plants increases it is measured as the primary productivity: the biomass produced by consumers and decomposers is measured as the secondary productivity.

Primary and secondary productivity are measurements used by scientists to determine an ecosystem's capacity to support life. An ecological crisis can take place when species or populations evolve in an unsustainable way. Sometimes the quality of the environment is lost after trauma, like a lack of rain or an increase in the temperature of a region. Other times, too much predation can ruin the balance of an ecosystem, as can be seen with overfishing by humans. Sometimes, too many organisms in one place will cause a poor living environment for others. There are different time scales for ecological crises ranging anywhere from a few months to millions of years. Extinction events can affect many species or just a few individual species. Human activities, like oil spills, can cause local crises as well as global crises like global warming. Even with a local crisis, the loss of a few species can have a disastrous effect on the survival of others in the food chain. A global crisis can result in a loss of nearly all species on Earth. The most common example of such a crisis is the extinction of the dinosaurs. Other examples of well-known ecological crises are the Permian-Triassic extinction event, the Cretaceous-Tertiary Event, Global Warming from the Greenhouse effect, the hole in the ozone layer, desertification and deforestation, and nuclear meltdowns like Chernobyl. Fortunately, nature always prevails and when species disappear, new species evolve.

Marine ecology deals with environment and life in marine waters. It involves the study of:

- Marine organisms and their habitat (physical support, light and transparency, nutrients, temperature, salinity and osmotic processes, gases, hydrostatic pressure, viscosity, circulation of water).
- Divisions of Marine environment (open sea-pelagic environment, epipelagic environment, mesopelagic zone, bathypelagic zone, abyssopelagic zone, and sea bottom- benthic environment).
- Distribution of marine life (planktons-floaters, Phytoplanktons, Zooplanktons, The nektons and the benthos).

Marine organisms and their habitat includes the:

- Physical support.
- Light and transparency.
- Nutrients.
- Temperature.
- Salinity and osmotic processes.
- Gases.
- Hydrostatic pressure.
- Viscosity.
- Circulation of water.

The Divisions of Marine environment includes the study of open sea:

- Pelagic environment.
- Epipelagic environment.
- Mesopelagic zone.
- Bathypelagic zone.
- Byssopelagic zone.
- Sea bottom- benthic environment.

The distribution of marine life includes the study of:

- Planktons-floaters.
- Phytoplanktons.
- Zooplanktons.
- Nektons and the Benthos.

Life on earth originated from oceanic waters. Creatures living on land need to deal with gravity, dehydration, great temperature variations, and many other factors which are normally not encountered in oceans. But the marine organisms are affected by the factors like:

- Depth of oceanic water.
- Change in salinity.
- Change in temperature.
- Change in turbidity.
- Other environmental factors.

Depth Zones in Oceans

The continental margin forms the part of the sea bed that borders all the land frontiers. It consists of three zones as Continental shelf, continental slope and Deep Ocean basins.

The continental shelf extends from the coast to a depth of 130 m. The width averages to about 75 kilometers. This shelf zone collects much of the sediments (deposits of sand and mud) that are carried by the rivers from land.

The continental slope begins at the outer edge of the shelf. The slope is much steeper than the shelf and plunges to great depths of 3.6 kilometers. The width ranges from 20 to 100 kilometres. Submarine canyons extend into these slopes. The canyon heads may form some deep sea fans and levees.

The area between continental slope and deep ocean floor is known as continental rise. The deep ocean basin is called as the abyssal plain. Abyssal hills, gyots, sea mounts and deep sea trenches are the physiographic features of the basin.

The deep ocean basins consist of deep-sea peaks, valleys, and plains which lie beyond the continental margin in the ocean basin. The mid-ocean ridges are the unique features of the ocean basins. Deep valleys also cut across the ridges in many places. Frequent volcanic activity is also expected in some valleys.

The two chief sources of deep-sea sediment are the land itself and marine life. Marine life sediment consists mainly of tiny shells and the remains of dead organisms of the plankton. When such matter makes up a large part of sediment, they are called as oozes.

Oceans	Area (Sq.km)	Average depth(m)
The Pacific	181	3,940
The Atlantic Ocean	94	3,580
The Indian Ocean	74	3,840

Salinity

Sea water contains much mineral salts to the extent of increasing its density. The oceans and seas put together contain about 5 x 10^{16} tons of salts. One cubic foot of average sea water contains 2.2 pounds of salt.

Salinity is generally reported in terms of ppt or parts per thousand (abbreviated o/oo), the number of pounds of salt per 1,000 pounds of water. It is affected by precipitation, evaporation and movement of water masses. It also varies from the surface zone to the deep ocean waters. Based on this factor, animals living in marine waters are classified into:

- Stenohaline animals: Restricted life with salinity.
- Euryhaline animals: Life that can be found at all levels of salinity.

The marine communities are adapted to live in high saline waters. The osmotic property of sea water is another problem of marine animals. The following are the ionic concentrations of oceanic waters:

Sea salts	Parts per Thousand
Chloride	19.3 o/oo
Sodium	10.7 o/oo
Sulfate	2.7 o/oo
Magnesium	1.3 o/oo
Calcium	0.4 o/oo
Potassium	0.4 o/oo
Bicarbonate	0.15 o/oo
Bromide	0.07 o/oo
Other	0.06 o/oo

Hence, the average salinity is 35 ppt. The saltiest water is in the Red Sea and also in the Persian Gulf, which have a salinity of about 40 ppt. The least salty seas are in the Polar Regions.

Temperature

The Temperature fluctuation is minimum in oceanic waters. Arctic waters are colder. Tropical waters are warmer. The variation is seasonal and diurnal. Ocean animals show a varied response to the temperatures.

There are two kinds of animals:

- Stenothermal animals: The ones which live within 20°C.

- Eurythermal animals: The animals which can withstand the wide range of temperature variations.

The cold water forms also show an increase their in sizes. The surface temperature of the oceanic waters vary from about -2 °C near the North and South poles to about 30 °C near the equator. In the Polar Regions, the surface sea water freezes. The western tropical Pacific has the warmest surface water. Ocean currents affect the surface temperature.

Ocean temperature also varies with depth. In general, the temperature falls as the depth increases. The warm surface waters extend to depths of about 150 metres in the tropics, and about 300 metres in the subtropics. Below the surface waters, the temperature drops rapidly, forming a layer called the thermocline. The thermocline varies in thickness, from about 300 metres to 910 metres. Below the thermocline, the water cools more slowly. Close to the deep-sea floor, the temperature of the ocean ranges between 1° and 4 °C.

Light and Photic Zones

Light is very essential for the marine life. For photosynthesis, heating, radiations and vision are related to this factor. It determines the plant and animal life and depends on the angle of incidence. Perpendicular rays go deeper into the oceans. Based on the intensity of light the water body of marine ecosystems can be differentiated into three zones. They are:

- Euphotic zone:
  - Extends up to 100 m depth.
  - Illuminated zone production zone.
  - Rich with phytoplankton, primary consumers and secondary consumers.
- Disphotic zone:
  - Lies between 100 and 200 m depth.
  - Weakly lighted.
  - Secondary consumers live with a few primary consumers, rare plant life.
- Aphotic zone:
  - Above 200m depth.
  - No light.
  - No producers.
  - No photosynthesis.
  - Secondary consumers.
  - Feed on other animals also.

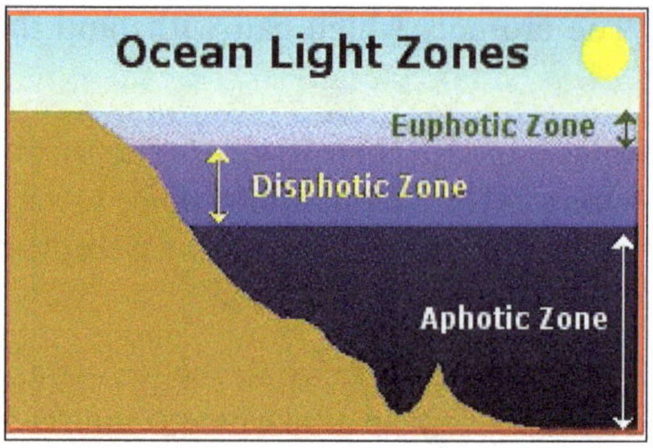

Euphotic Zone is the uppermost layer of the world's oceans. Euphotic means "well lit" in Greek. In this zone, there is sun light for photosynthesis. Hence, many plants and photosynthetic organisms live in this zone and food is abundant. These primary producers (also called as autotrophs) are the first link in the food chain in the oceans. Due to this, many animals also live in this zone. In fact, most of the marine life live in the ocean is found in this zone. In the euphotic zone, photosynthesizers include:

- Free-floating algae- often called seaweeds:
  - Red algae (Rhodophyta).
  - Green algae (Chlorophyta).
  - Brown algae (Phaeophyta).
- Phytoplankton: Tiny, one-celled photosynthetic plankton like diatoms, dinoflagellates, and Coccolithophorids.
- Flowering plants (angiosperms): Submerged Sea grasses.

Euphotic zone animals include most ocean fish (including sharks and rays), man-o'-war, jellyfish, sea turtles, seals, corals, and zooplankton. Some bottom-dwellers also live in the euphotic zone. Some of the animals have countershading characteristics. This adaptation helps to camouflage the organism, hiding it from predators and allowing it to sneak up on prey. Most of the sharks are counter shaded.

Twilight Zone

The middle layer of the world's oceans receives only faint, filtered sunlight during the daytime. This is because the seawater absorbs the sunlight. Disphotic means "poorly lit" in Greek. This zone appears deep blue to black in color. The depth of this zone depends on the clarity or murkiness of the water. The disphotic zone extends to about 1,000 m. This is where the aphotic zone begins. In the disphotic zone, there is enough light to see only during the day, but not enough light for photosynthesis to take place.

Hence, no plants can live in this zone. The amount of light decreases with depth. Because of this, food is not abundant in this zone. The water in the disphotic zone is cold and decreases with depth. The pressure is high - it can be up to 1,470 psi (pounds per square inch) and increases with depth. The amount of dissolved oxygen in the water is also less than in this zone. The animals that live in the disphotic zone are adapted to live in near darkness, cold water and high pressure. Many of the animals have large eyes, helping them to see in the nearly dark waters. Many have large teeth and jaws.

Bioluminescence is a characteristics feature of the animals in this zone. Many of these animals can make their own light. Some bioluminescent animals have special organs that produce light by a chemical reaction; other bioluminescent animals have glowing bacteria that live on them. Most bioluminescent organs called photophores give off an eerie blue-green light. Counter-illumination is a method of camouflage.

Light produced on the underside of bioluminescent animals can help the animal "disappear" from predators that are below it. Animals in the disphotic zone are filter feeders, grazers, and predators. Some animals migrate vertically (up and down) in order to feed upon the abundant life in the bright zone above them.

Animals begin to decay as soon as they die. In addition, animals give off some waste products also. These waste products--and the dead animals that are not eaten--are broken down into mineral salts by bacteria. Rising currents carry the minerals to the surface.

Ocean Water Pressure

The difference in pressure (pgh) comes due to the:

- Thickness of water column (h).
- Density of water(p).
- Acceleration due to gravity (g).

This has a great influence over the life and deposition, and solubility of $CaCO_3$ mechanisms. There are two classes:

- Eurybathic- Animals live with great vertical range (Eg. Natica).
- Stenobathic- Narrow range (E.g. Turris).

Other factors: The other abiotic factors which have much impact on marine ecosystems are tides, currents and waves.

The currents are of three types such as:

- Density currents.
- Tidal currents.
- Wind currents.

Two types of circulation create the currents in the ocean. They are:

- Wind-driven circulation.
- Thermohaline circulation.

Wind-driven circulation results from the wind blowing on the ocean surface. The currents generally flow horizontally--that is, parallel to the earth's surface. Wind-driven currents move in enormous circular patterns called gyres.

Thermohaline circulation produces great vertical currents that flow from the surface to the ocean bottom and back. The currents largely result from differences in water temperature and salinity. One of the most dramatic forms of weather occurs over the oceans. Tropical Cyclones including "hurricanes" and tornadoes are expected to play a significant role in the marine systems.

Marine Life

Marine life ranges in size from microscopic one-celled organisms to the blue whale, which may measure up to 30 metres long. Ocean plants and plantlike organisms use sunlight and the minerals in the water to grow. Sea animals eat these organisms and one another.

Based on the variations in physical condition and the presence of certain specific kinds of animals and plants, the Marine environment is broadly classified into:

- Pelagic zone: Open Sea -entire sea water above ocean floor -encompasses a neritic zone and an oceanic zone.
- Benthic zone: Ocean (bottom)-encompasses the littoral and deep sea zones.

All ocean life can be divided into three groups. These groups are:

- The plankton.
- The nekton.
- The benthos.

Planktons are wandering forms. Some rare freely floating plants. Some are organisms. The marine planktons include snails, crab, starfish, larval fish, barnacles, worms and sea cucumbers. The phytoplanktons include flagellates, diatoms, coccoplithophores and dinoflagellates.

The zooplankton consists of many kinds of animals, ranging from one-celled organisms to jellyfish up to 1.8 metres wide. The animals live in the near surface and deep waters of the ocean. Many organisms settle at the sea floor or attach themselves to it and become a part of the benthos.

The nekton consists of animals that can swim freely in the sea. Whales belong to this category. Sea snakes, turtle and sea otters are notable animals. They are strong swimmers including fishes, squids, and marine mammals. Most nektons live near the sea surface, where the food is plenty. But many other animals of the nekton live in the deep ocean.

Fish are the most important animals among all nektons. About 13,300 kinds of fish live in the oceans. They differ greatly in their sizes and shapes. The largest fish, the whale shark, measures as much as 12 metres long.

Squids are free-swimming molluscs (animals with soft, boneless bodies) that have 10 arms. Nektonic mammals include dugongs, manatees, porpoises, and whales. All of them remain in the ocean all through their lives. Other marine mammals, such as sea lions, sea otters, seals, and walruses, spend time on land. Based on the populations and communities present in the marine ecosystems, the following zones have been recognized:

- Neritic (or) Eulittoral zone: This is the shallow water zone belonging to the Inter tidal area on the continental shelf.
- Oceanic zone: Open sea beyond the continental shelf extending upto the depth of effective light penetration.
- Bathyal zone or archbenthic zone: Beyond the depth of light penetration from 200 -1000m depth.
- Abyssal zone: Beyond 1200 m depth- dark zone.

Littoral Zone refers to the inter-tidal area where the land and sea meet, between the high and low tide zones. This complex marine ecosystem is found along coastlines worldwide. It is rich in nutrients and oxygen and is home to a variety of organisms. This inhospitable environment is washed by the tides each day. Hence, the organisms that live here are adapted to huge daily changes in four factors as:

- Moisture.
- Temperature.
- Turbulence (from the water).
- Salinity.

Moisture: The littoral zone is covered with salt water at high tides, and it is exposed to the air at low tides; the height of the tide exposes the land to this daily tide cycle. Organisms must be adapted to both very wet and very dry conditions.

Water Movement: The turbulence of the water is a limiting factor. Rough waves can dislodge or carry away poorly-adapted organisms. Many inter-tidal animals burrow into the sand (like clams), and live under the rocks, or attach themselves to rocks (like barnacles and mussels).

Animals that live in the littoral zone have a wide variety of predators who eat them. When the tide is in, littoral organisms are preyed upon by sea animals (like fish). When the tide is out, they are preyed upon by land animals, like foxes and people. Birds (like gulls) and marine mammals (like walruses) also prey on inter-tidal organisms extensively.

Threats to Marine Biology

Marine ecosystems and aquatic habitats face a number of threats from humans. Serious conservation attention and efforts should be drawn and directed towards restoration of fragmented marine habitats and estuarine ecosystems. Threats are manifold: i.e., primarily from overexploitation of marine resources, overfishing, climate warming, sewage disposal, industrial chemical discharge, oil spills, invasive species, and dredging. The Corals are particularly fragile animals and very sensitive to changes in temperature and salinity. Coral reefs are coastal habitats build by corals that are also one of the most productive habitats within the marine ecosystems, but they are however, at great danger of going extinct from any drastic changes in sea surface temperature and salinity. Ecosystem services provided by coastal areas helps to sustain coastal people whose livelihood depends on fishing and fish products, aquaculture and exploitation of biologically diverse resources that the seas and the oceans provide them.

Marine Pollution

Oceans are the largest water bodies on the planet Earth. Over the last few decades, excessive human activities have severely affected marine life on the Earth's oceans. Ocean pollution, also known as marine pollution, is the spreading of harmful substances such as oil, plastic, industrial and agricultural waste and chemical particles into the ocean.

Since oceans provide the home to a wide variety of marine animals and plants, it is the responsibility of every citizen to play his or her part in making these oceans clean so that marine species can thrive for a longer period of time.

Mining for materials such as copper and gold is a major source of contamination in the ocean. For example, copper is a major source of pollutants in the ocean and can interfere with the life cycles of numerous marine organisms and life.

Various Causes of Ocean Pollution

There are various ways in which pollution enters the ocean. Some of them are:

Sewage

Pollution can enter the ocean directly. Sewage or polluting substances flow through sewage, rivers, or drainages directly into the ocean. This is often how minerals and substances from mining camps find their way into the ocean.

The release of other chemical nutrients into the ocean's ecosystem leads to reductions in oxygen levels, the decay of plant life, and a severe decline in the quality of the seawater itself. As a result, all levels of oceanic life, plants and animals, are highly affected.

Toxic Chemicals from Industries

Industrial and agricultural waste is another most common form of wastes that is directly discharged into the oceans, resulting in ocean pollution. The dumping of toxic liquids in the ocean directly affects marine life as they are considered hazardous, and secondly, they raise the temperature of the ocean, a phenomenon known as thermal pollution, as the temperature of these liquids is quite high. Animals and plants that cannot survive at higher temperatures eventually perish.

A plastic bottle would require about 450 years to decompose completely. In the meanwhile, it would remain inside the ocean, releasing toxins that would poison the marine life.

Land Runoff

Land runoff is another source of pollution in the ocean. This occurs when water infiltrates the soil to its maximum extent, and the excess water from rain, flooding or melting flows over the land and into the ocean. Often, this water picks up man-made, harmful contaminants that pollute the ocean, including fertilizers, petroleum, pesticides and other forms of soil contaminants. Fertilizers and waste from land animals and humans can be hugely detrimental to the ocean by creating dead zones.

Large Scale Oil Spills

Ship pollution is a huge source of ocean pollution, the most devastating effect of which is oil spills. Crude oil lasts for years in the sea and is extremely toxic to marine life, often suffocating marine animals to death once it entraps them. Crude oil is also extremely difficult to clean up, unfortunately, meaning that when it is split, it is usually there to stay.

Besides, many ships lose thousands of crates each year due to storms, emergencies, and accidents. This causes noise pollution (excessive, unexpected noise that interrupts the balance of life, most often caused by modes of transportation), excessive algae, and ballast water.

Often, other species can also invade an ecosystem and do harm to it by interrupting the life cycles of other organisms, causing a clash of nature that has already been damaged by the overflow of pollution.

Ocean Mining

Ocean mining in the deep sea is yet another source of ocean pollution. Ocean mining sites drilling for silver, gold, copper, cobalt, and zinc create sulfide deposits up to three and a half thousand meters down into the ocean.

While we yet to have the gathering of scientific evidence to fully explain the harsh environmental impacts of deep-sea mining, we do have a general idea that deep-sea mining causes damage to the lowest levels of the ocean and increases the toxicity of the region. This permanent damage dealt also causes leaking, corrosion and oil spills that only drastically further hinder the ecosystem of the region.

Littering

Pollution from the atmosphere is, believe it or not, a huge source of ocean pollution. This occurs when far inland objects are blown by the wind over long distances and end up in the ocean.

These objects can be anything from natural things like dust and sand to man-made objects such as debris and trash. Most debris, especially plastic debris, cannot decompose and remains suspended in the ocean's current for years.

Animals can become snagged on the plastic or mistake it for food, slowly killing them over a long period of time. Animals who are most often the victims of plastic debris include turtles, dolphins, fish, sharks, crabs, sea birds, and crocodiles.

Also, the temperature of the ocean is highly affected by carbon dioxide and climate changes, which primarily impacts the ecosystems and fish communities that live in the ocean. In particular, the rising levels of Co_2 acidify the ocean in the form of acid rain.

Even though the ocean can absorb carbon dioxide that originates from the atmosphere, the carbon dioxide levels are steadily increasing, and the ocean's absorbing mechanisms, due to the rising of the ocean's temperatures, are unable to keep up with the pace.

Devastating Effects of Ocean Pollution

Effect of Toxic Wastes on Marine Animals

The oil spill is dangerous to marine life in several ways. The oil spilled in the ocean could get on to the gills and feathers of marine animals, which makes it difficult for them to move or fly properly or feed their children. The long term effect on marine life can include cancer, failure in the reproductive system, behavioral changes, and even death.

Disruption to the Cycle of Coral Reefs

Oil spill floats on the surface of the water and prevents sunlight from reaching to marine plants and affects the process of photosynthesis. Skin irritation, eye irritation, lung and liver problems can impact marine life over a long period of time.

Depletes Oxygen Content in Water

Most of the debris in the ocean does not decompose and remain in the ocean for years. It uses oxygen as it degrades. As a result of this, oxygen levels go down. When oxygen levels go down, the chances of survival of marine animals like whales, turtles, sharks, dolphins, penguins for a long time also go down.

Failure in the Reproductive System of Sea Animals

Industrial and agricultural wastes include various poisonous chemicals that are considered hazardous for marine life. Chemicals from pesticides can accumulate in the fatty tissue of animals, leading to failure in their reproductive system.

Effect on Food Chain

Chemicals used in industries and agriculture get washed into the rivers and from there are carried into the oceans. These chemicals do not get dissolved and sink at the bottom of the ocean. Small animals ingest these chemicals and are later eaten by large animals, which then affects the whole food chain.

Affect Human Health

Animals from impacted food chain are then eaten by humans, which affects their health as toxins from these contaminated animals get deposited in the tissues of people and can lead to cancer, birth defects or long term health problems.

Solutions to Ocean Pollution

Ocean pollution is a raging problem that needs to be solved as early as possible. With the world's oceans getting polluted, the marine ecosystems are getting severely disturbed.

Not only that, but the world's water reserves are also becoming more and more limited. In these situations, it is essential that we find ways of solving ocean pollution. Some of the ways are as follows:

Reducing the use of Plastic Products

Believe it or not, plastic wastes form the largest portion of ocean pollutants. Out of the 260 million tons of plastic produced each year globally, approximately 10% ends up in the oceans.

These plastics then take thousands of years to decompose, during which time it posses a terrible threat to life in the oceans. Thus, reduction in the use of plastic products could help in significantly reducing the rates of ocean pollution.

Use Reusable Bottles and Cutlery

Use and throw bottles and cutlery, too, are an immense contributor to ocean pollution. We must not forget that most of the non-reusable bottles and cutlery are made out of either plastic or Styrofoam. Both these materials require a few hundred or thousand years to decompose.

A plastic bottle would require about 450 years to decompose completely. In the meanwhile, it would remain inside the ocean, releasing toxins that would poison the marine life.

Recycle Whatever you can

The 3 R's are definitely the savior of ocean health. With the increase in population, it is true that the waste produced would also increase significantly. As finding dumping grounds become more and more difficult, a lot of the wastes find their way to the ocean beds. These wastes that are dumped in the ocean remain there for a long time, affecting marine life.

One of the useful ways to reduce waste production is recycling. Before throwing things away, we could check if something is recyclable. Then the products that can be recycled can be taken to the recycling center nearby.

Stop Littering the Beach and Start Cleaning

The beach is undoubtedly one of the loveliest places to visit and hang out in. As a result of this, there is a lot of littering around as well. The first and foremost thing that really needs to stop is this. Abundant dustbins should be placed on the beaches.

If someone is seen littering, they must be stopped immediately, and certain strict rules and regulations must be followed at all times. Also, if the beach seems untidy, we could pick up the wastes and throw it in the dustbin.

Reducing the use of Chemical Fertilizers

Runoffs, too, pollute the oceans immensely. To prevent this, the use of chemical fertilizers must also be controlled and regulated. It must be remembered that excess use of chemical fertilizers harms not only the soil but also the water bodies nearby and, ultimately, the ocean. These runoffs are so toxic that they can very easily kill marine life.

Reducing the Energy Use

As the population increases, our demand for energy increases too. A major chunk of this energy is produced from petroleum. This petroleum is obtained by the drilling of the ocean beds. Any spillage during the procedure could harm marine life. Not only has that, but the procedure itself also violated the marine ecological balance. So by controlling our energy use, we could also solve the problem of ocean pollution.

Oil Pollution Toxicity to Marine Life

Petroleum marine fuel spills, which result from damage, transportation accidents and various other industrial and mining activities, are classified as hazardous waste. They are considered to be the most frequent organic pollutants of aquatic ecosystems.

In recent years, there have been numerous studies regarding the levels of contamination of the seawater by hydrocarbons. The majority of these studies were conducted following the Gulf War of 1991 and after, the BP Deepwater Horizon (DWH) oil spill on 20th April, 2010.

Major Oil Spills Related to Marine Environments

Saltwater bodies are referred to as 'marine environments', with the oceans covering about 70% of the earth. Statistics estimate that 3.2 million tonnes of oil per year are released from all sources into the environment. The majority of this oil is due to general shipping and industrial activities. During the Iran-Iraq war, approximately 2 million barrels of oil were discharged into the Arabian Gulf sea water. These included 1.5 million barrels from the Nawruz blow-out in 1983. Following the Gulf War in 1991, between 4 and 8 million barrels (1,000 tonnes = 7,500 barrels) were released into the Gulf and the Kuwaiti Desert, making this the largest oil spill in history at that time. When compared to other major spills, the size of this spill attracted global attention. For example, the Amoco Cadiz off the coast of Brittany (France), spilling 200,000 tonnes (1.5 million barrels); the Torrey Canyon, Braer, Sea Empress and the super tanker Breaf off the coast of Shetland (UK) in 1993, spilling a maximum of 84,000 tonnes (607,300 barrels); the Exxon Valdez in Prince William

Sound or Alaska (US), which was approximately 36,224 tonnes (261,904 barrels). Of the recorded oil spills, more than 80% were less than 1,000 tonnes (7,500 barrels) and only 5% were greater than 10,000 tonnes. An accepted average sample size of an oil spill is about 700 tonnes (5,061 barrels).

Table: Number of oil spills over 7 tonnes from 1970 to 2013.

Year	7-700 tonnes	>700 tonnes	Quantity Spilled x 10^3 tonnes
1970-1974	189	125	1114
1975-1979	342	117	2012
1980-1984	221	41	570
1985-1989	124	48	513
1990-1994	165	48	907
1995-1999	108	25	194
2000-2009	182	-	213
2010-2013	28	-	22

During the 1990s, the average number of large spills per year was about a third of the amount that was witnessed during the 1970s. When looking at the quantities of oil spilled, it should be noted that 1, 133, 000, 213, 000 tonnes of oil were lost in the 1990s and 2000s, respectively. In a four-year period, 2010-2013, there were 22,000 tonnes of oil lost. Table below shows a brief summary of the top 22 major oil spills since 1967.

Table: Top 22 major oil spills since 1967.

Ship name	Year	Location	Spill (10^3) tonnes
Torrey Canyon	1967		119
Sea Star	1972		115
Jakob Maersk	1975		88
Urquiola	1976		100
Hawaiian Patriot	1977		95
Amoco Cadiz	1978		223
Atlantic Empress	1979		287
Independenta	1979		94
Irenes Serenade	1980		100
Castillo de Bellver	1983		252
Novo	1985		70
Odyssey	1988		132
Khark 5	1989		70
Exxon Valdez	1989		37
ABT Summer	1991		260
Haven	1991		144
Aegean Sea	1992		74

Katina P	1992	Off Maputo, Mozambique	67
Braer	1993	Shetland Islands, UK	85
Sea Empress	1996	Milford Haven, UK	72
Prestige	2002	Off Galicia, Spain	63
Hebel Spirit	2007	South Korea	11

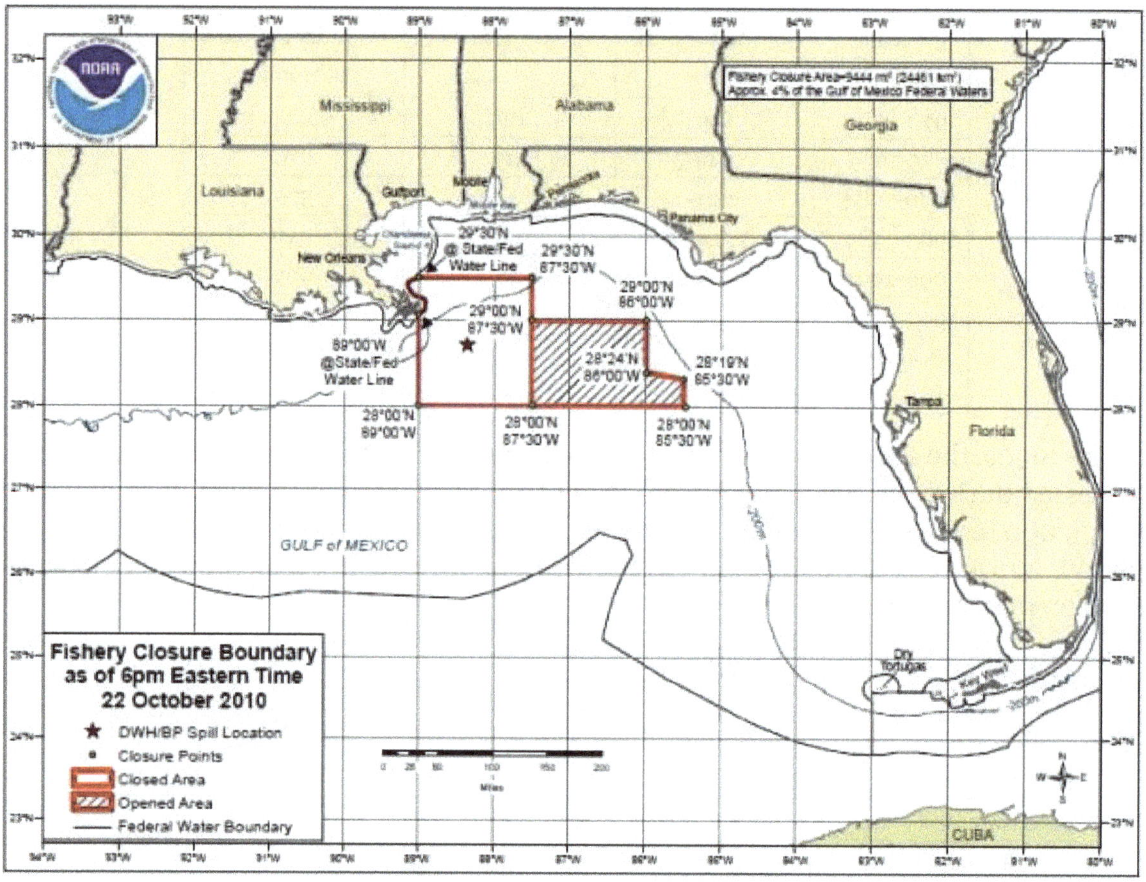

Figure: The Site of the Gulf Spill: Closed Areas as of 22nd October, 2010.

Fate of Oil Spills in the Marine Habitats

After oil is spilled at sea and with the effect of wind and water current, the oil spreads out and moves on the water surface as a slick a few millimeters thick. At the same time, it undergoes a series of chemical and physical changes. These processes are collectively termed 'weathering'. Weathering causes the spilled oil to break down and become heavier than water. Some of these processes, like the natural dispersion of oil into water, lead to the removal of the oil from the sea surface and facilitate its natural breakdown in the marine environment. Others, particularly the formation of water-in-oil emulsions, cause the oil to become more persistent and remain at sea or on the shoreline for prolonged periods of time. The speed and relative importance of these processes depend on a number of factors. These include the quantity spilled, the oil's initial physical and chemical characteristics, weather and sea conditions and whether the oil remains at sea or is washed ashore. Ultimately, the marine environment usually eliminates spilled oil through the long-term process of biodegradation. These include weathering, evaporation, oxidation and biodegradation.

Impact of Oil Spills on Marine Organisms

Ultimately, the impact of oil on marine organisms depends on the fate of the oil. When oil is present in the environment, it is either dispersed in the top layer of the water (littoral zone) or remains on the surface and, consequently, on the coastal areas. If the oil is not dispersed, it remains on the surface. In this case, currents bring the oil towards coastal areas which harms coastal organisms like invertebrates, mammals and birds. However, if the oil is dispersed, organisms, such as fish, plankton and larvae, are immediately subjected to oil toxicity.

Impact of Oil Spills on Planktonic Organisms

Zooplankton is a particularly important food resource, especially for baleen whales. It can influence or control the primary productivity by top-down effects in return. Its population dynamic change can influence the biomass of other marine animals like fish by bottom-up effects. Some zooplankton, such as copepods, euphausiids and mysids, assimilate hydrocarbons directly from seawater and by ingesting oil droplets and oil contaminated food. The ingestion of oil by these organisms often causes mortality; while surviving organisms often show developmental and reproductive abnormalities. The impact of petroleum pollution on marine plankton has been a great cause for concern.

The reports regarding the toxic effects of oil water accommodated fraction (WAF) on marine phytoplankton, zooplankton and the early life stages of animals. Generally, oil WAF toxicity enhances with increasing carbonic chain length and benzene ring number. It also suggested future study points to further promote the quantified evaluation of the damage by oil pollution to marine ecology.

For the oil WAF, reported that plankton are capable of accumulating PAH due to their great lipophilic abilities. They, therefore, stimulate various harmful effects. The investigators reported that marine plankton is highly sensitive to the petroleum WAF, as the order of median effective/lethal concentration is as low as lg/L or mg/L. Examinations of the toxicity effect of 10 polycyclic aromatic hydrocarbons associated with the Prestige fuel oil spill on adult copepods (Oithona davisae) revealed that the PAHs had narcotic effects on these organisms.

Impact of Oil Spills on Benthic Organisms and Invertebrates

Benthic invertebrates and higher forms, such as the sand eel and Ammodvtes americanus (a main food resource of Atlantic humpback whales), may accumulate petroleum hydrocarbons from water, contaminated sediments and food. Thus, these whales are adversely affected.

Bivalve molluscs tend to accumulate petroleum hydrocarbons to higher concentrations and retain them longer than other taxa. This is essentially due to the lack of a mixed function oxygenase (MFO) system that makes them unable to metabolize the compounds to execrable polar metabolites. Thus, they are likely to transfer them to their predators. Marine mammals that rely heavily on bivalve molluscs for food, such as the walrus and otter, share a higher risk of ingesting petroleum hydrocarbons. Benthic amphipods are quite sensitive to spilled oil. They are among the first marine animals killed and the slowest to recover. However, most marine crustaceans have a well-developed MFO system. As a result, they are able to metabolize and excrete accumulated hydrocarbons quite rapidly.

In the intertidal habitat, the biological recovery of the exposed shores is faster than the sheltered shores. This is because strong wave action promotes the removal of contamination and the animals and plants of exposed shores tend to be severely affected. Thus, they are better able to re-colonize an impacted shore quickly. Sublittoral habitats are generally contaminated by sedimentation of oiled particulate material and clean-up for these habitats is not practiced. As a result, the recovery of subtidal communities impacted by oil spills usually takes a longer time. For example, the Abra alba bivalve sand community in the Bay of Morlaix, Brittany was severely affected by the Amoco Cadiz oil spill (1978). After the spill in 1978, the biomass values for the sand community immediately fell. However, within two years, they recovered to pre-spill levels. Productivity also showed similar trends.

Invertebrate populations, such as the amphipod sand hopper, Ampelisca, are quite sensitive to oil pollution and, for various reasons, are slow to re-populate and 'recover'. For example, the initial impact of the spill in the Bay of Morlaix, following the Amoco Cadiz spill in 1978, was to kill off populations of the amphipod sand hopper, Ampelisca, which dominated the community. Although the sediment was rapidly purged of the contaminating oil, it has been noted that Ampelisca was back to its pre-spill population density after 10 years. The standing crop biomass and productivity was restored much more rapidly as the ecological niches had been occupied by other opportunists like the bivalve Abra and the worm Nephtys. These had quickly filled the place left by the Ampelisca.

In other studies on oil spill accidents, it has been shown that six months after the Prestige oil spill and clean-up campaign, invertebrate populations of the exposed sandy beaches, notably the isopod Eurydice, the spionid polychaete Scoleleoiussquamata, nemerteans and Diptera, were significantly reduced. Furthermore, their abundance inversely related with the oil pollution gradient. The number of taxa was reduced but not the diversity values. The only clam on these sandy beaches was Donax. Before the spill and clean-up, it occurred in six beaches but afterwards, only in one. Upper dry sand communities were particularly reduced due to both oil toxicity and extensive beach grooming that also removed seaweed wrack.

Marine invertebrates are unique living organisms in different aspects. They have been explored as a model for several biological markers as a result of oil spill pollution. The marine invertebrate mussels (Mytilus edulis) exposed in situ to the oil that came ashore after the wreck of the 'Erika' tanker on the Brittany (France) coast in December, 1999. The mussel response was assessed using a set of biomarkers (acetylcholinesterase (AChE), glutathione S-transferase (GST), catalase (CAT), malondialdehyde (MDA) and deoxyribonucleic acid (DNA) adducts, related to the metabolism of the organic contaminants.

The results of a series of validation tests revealed that there were no significant reductions in the GST or CAT levels. Six months immediately following the accident, observations indicated that the DNA adducts and MDA levels were high and the levels of AChE were significantly lower during the first year of the survey. This suggested a general stress. A simple multivariate graphic method - the integrated biomarker response index - was used to combine four of the five validated biomarkers and quantify the degree of impact on mussels at different sites. The results indicated that mussel populations were affected by the oil spill during only the first year after the accident.

Impact of Oil Spills on Coral Reefs

In addition, recreational attractions for divers, coral reefs are considered to be important constituents of marine ecosystems. This is because they are important nurseries for shrimp, fish and other animals. The aquatic organisms that live within and around the coral reefs are at risk of exposure to the toxic substances within oil, as well as smothering. They are rapidly deteriorating because of a variety of environmental and anthropogenic pressures. Thus, they are suffering significant changes in diversity, species abundance and habitat structure worldwide.

Oil dispersants are potentially harmful to marine life including coral reefs. In a study using coral nubbins in coral reef ecotoxicology testing, found that dispersed oil and oil dispersants are harmful to soft and hard coral species at early life stages. The investigators also employed a 'nubbin assay' on more than 10, 000 coral fragments to evaluate the short- and long-term impacts of dispersed oil fractions (DOFs) from six commercial dispersants (Slickgone, Petrotech, Inipol, Biorieco, Emulgal and Dispolen) and the dispersants and watersoluble-fractions (WSFs) of Egyptian crude oil on two Indo-Pacific branching coral species, Stylophora pistillata and Pocillopora damicornis. They found that the dispersant concentrations recommended by the manufacturer were highly toxic and resulted in mortality of all nubbins. The dispersed oil and the dispersants were significantly more toxic than the crude oil WSFs. As corals are very sensitive to oil detergents and dispersed oil, the results of these assays indicated the negative and harmful effect of using any oil dispersant in coral reefs and in the area closely around them. These dispersants were rated based on their eco-toxicological impacts on the corals. Observations indicated a scale from the least to the most harmful agent, as follows: Slickgone > Petrotech > Inipol > Biorieco > Emulgal > Dispolen.

Impact of Oil Spills on Fish

Due to the well-developed hepatic mixed function oxygenase (MFO) system, in addition to the reactivity of the metabolites that would not be released in a toxic form during digestion and absorption, most fish, even in heavily oil-contaminated environments, do not accumulate and retain high concentrations of petroleum hydrocarbons. Thus, they are not likely to transfer them to predators.

Generally, marine carnivores are inefficient assimilators of petroleum compounds in food. For this reason, and because all prey species are able to release hydrocarbons from their tissues, the marine food chain biomagnification does not occur. Thus, there is an indirect correlation between a marine mammal's trophic level and the concentration of residues that it might consume. In fact, as top carnivores that feed on large pelagic fish and seals, polar bears and killer whales are less likely to be exposed to petroleum in their food than other species, such as walrus and baleen whales, which feed on zooplankton and benthic invertebrates.

Chronic exposure of juvenile pink salmon (Oncorhynchus gorbuscha) to Alaska North slope crude oil resulted in a variety of responses. These included melanosis, erratic swimming, loss of equilibrium, reduced mobility and startled response. Interestingly, the effects were not enhanced by the phototoxicity from UV irradiation, presumably of the highly pigmented nature of the fish's skin (in contrast to phototoxicity in translucent early life stages of marine invertebrates). When they examined the phototoxicity of the water accommodated fractions of weathered crude oil, they found that pink salmon may be at less risk of photoenhanced toxicity, compared to the more translucent early-life stages of several other Alaskan species.

Impact of Oil Spills on Seabirds

As one of the major routes of exposure, physical contact usually affects birds. For example, thousands of African penguins (Spheniscus demerus) were oiled following the 2000 Treasure oil spill in South Africa.

An evaluation of the impact of oil spills on seabirds has not been fully appreciated during incidents, despite pressure from the public concern, media and other interested parties for precise and up-to-date information on the damage. Consequently, the approximate numbers of seabird casualties involved in many major spills have only been estimated, while impacts at the population level have been difficult to determine. Natural variation and the huge range of factors that influence bird population statistics make it difficult to assess the impact of oil spill on sea birds.

In their efforts to assess the impact of oil spills on seabirds in Europe and in North America, the investigators reported that there are two inter-linked aspects to dealing with oiled seabirds during major spills. The first is a relatively expensive and logistically complicated process. This involves birds found alive and the humanitarian efforts made to clean, rehabilitate and successfully release them into the wild. The second involves assessing the likely impact of the spill on the populations of those species affected.

The relation between the size of an oil spill and the number of seabird casualties is not directly proportional. Moreover, estimates of the number of seabird deaths from oil slicks are highly speculative. This is because an unknown number of oiled birds may die at sea and not reach the coast. For example, following the Exxon Valdez oil spill, over 35,000 seabird carcasses were recovered in the northern Gulf of Alaska. However, after the Braer accident, even though the Braer spill (85,000 tonnes) was almost two and a half times as large as that of the Exxon Valdez, only 1,500 dead birds were counted.

Researchers argued that the rapid recovery of the murre breeding colonies in Alaska and the number of dead birds might be lower than was estimated. They also suggested that breeding pairs may have been replaced by younger birds that came in from the open sea. On the other hand, argued that the impacts of the spill may have been diffused over a large area, permitting local recovery and making it difficult to detect any changes in local abundance or habitat occupancy. Nine years after the Exxon Valdez oil spill, in their long-term effects observation, claimed that the populations of most bird species have not recovered and others still show potential population effects.

During the winter season of 2003, approximately 64,000 tonnes of Prestige heavy fuel oil spilled off Galacia, Spain. As a result, 10 % of the European shag (cormorants, Phalacrocorax aristotellid) were killed. In turn, this resulted in a 50% reduction in the 2003 breeding success of oiled colonies, compared to the unoiled colonies. It was found that the kidneys of 32 oiled shag had extensive tissue damage and haemorrhaging. They further found that the oiled shag were heavily infected with an eimeriorin Apicomplex coccidian, which under normal circumstances is not pathogenic.

Table: Comparison of the number of dead seabirds recovered following the ExxonValdez and Braer oil spills.

Species group	Alaskan spill	Shetland spill
Sea ducks (eiders, etc.)	1440	167
Mergansers	121	1

Loons	395	14
Grebes	462	0
Heron	1	3
Geese/swans	9	0
Gulls	696	74
Kittiwakes	1125	133
Cormorants/shags	836	864
Shearwaters	3399	0
Fulmars	870	31
Guillemots/murres	20562	220
Other auks	2174	29
Bald eagles	125	NA
Other birds	3152	0
Total	35467	1536

Impact of Oil Spills on Marine Mammals

Marine mammals include bottlenose dolphins, fins, humpbacks, rights, sei whales, sperm whales, manatees, cetaceans, seals, sea otters and pinnipeds. The physical contact of oil with furred mammals usually affects these animals because they rely on their outer coats for buoyancy and warmth. Consequently, these animals often succumb to hypothermia, drowning and smothering when oil flattens and adheres to the outer layer.

As part of their activities, all marine mammals spend a considerable amount of time at the surface. Here, they swim, breathe, feed or rest. Thus, the possibility of their contact with a surface slick, water-in-oil emulsion, or tar balls, is high. In heavy pelage marine mammals, such as fur seals, sea otters and polar bears, this contact may lead to fouling. Polar bears and otters groom themselves regularly as a means of maintaining the insulating properties of the fur and may, thereby, ingest oil. Animals with smooth surfaces or relatively little to no pelage, such as whales, dolphins, manatees and most seals, have an advantage as oil would have fewer tendencies to adhere to their surface.

Some baleen whales are skim-feeders, i.e., they eat at the surface. When in an area of slick or tar balls, this behaviour can lead to foul in the feeding apparatus. Tarry residues, in particular, can coat the baleen plates.

Animals, such as narwhals, belugas, ringed seals, walruses and polar bears in Polar Regions, spend most of their time at the ice edge in leads, polynyas and breathing holes. This is where spilled oil tends to accumulate.

Oil that contaminates a shore is likely to severely affect pinnipeds. Pinnipeds require such areas for nursery and, to a lesser extent, otters and bears. Some of the oil is eventually returned in subtidal sediments, where it may transfer to gray whales, walrus and some seals. Such species feed heavily on benthic animals.

When marine mammals encounter fresh oil, they are likely to inhale volatile hydrocarbons evaporating from the surface slick. These volatile fractions contain toxic monoaromatic hydrocarbons (benzene, toluene and xylenes) and low molecular weight aliphatics with anaesthetic properties.

The inhalation of these volatile hydrocarbon compounds is potentially harmful. The inhalation of concentrated petroleum vapours can cause the inflammation of and damage to the mucus membranes of airways, lung congestion or even pneumonia. Volatile benzene and toluene, which can be inhaled, can be transferred rapidly from the bloodstream into the lungs. Furthermore, they can accumulate from the blood into the brain and liver, causing neurological disorders and liver damage.

Marine mammals are probably poor accumulators to oil directly from the solution or dispersion in the water column. This is because the skin of cetaceans is relatively impermeable to oil. Additionally, most marine mammals do not drink large volumes of seawater. Thus, a significant accumulation of hydrocarbons by this route is unlikely to occur.

There is an extensive and diverse database regarding subjects that deal with effects of oil on marine mammals and those aspects of an animal's life history vulnerable to exposure of spilled oil. This database is summarized in a publication by the Department of Interior/Minerals Management Service (MMS 88-0049)/Atlantic OCS region, Canada. At the same time the physiological and toxicological effects of oil on each of the marine mammal groups.

'Physiologic and Toxicologic Effects on Pinnipeds', indicated that pinnipeds are inappropriately sensitive to the harmful properties of oil. Incidental ingestion during feeding, exposure to vapour concentrations and surface contamination with relatively fresh oil does not appear to cause a disaster. However, Pinnipeds trapped near the source of a spill or those which are forced to emerge in heavy accumulations of oil in leads and around rookeries exhibit the most severe effects. Experimental studies by the same investigators on fur seals indicated that surface fouling decreases the insulation value of the pelt. This can potentially lead to thermal and energetic stress. Furthermore, the sensitivity to the effects of oil exposure may be high in species and groups that are compromised by pre-existing disease, or stressed by pressures of an unfavourable habitat, intra-specific competition or unusual environmental conditions.

Impact of Oil Spills on Marine Plants

In several aspects, aquatic plants are important to the functioning of ecosystems. These include the fact that they are oxygen producers, their ability to sequester carbon and for their base position in aquatic food chains. In addition, they serve as nursery, feeding and breeding habitats for a variety of animal and plant species, including recreationally and commercially important fish. Plants and animals are affected by the oil in which they come into contact with as a result of an oil spill.

The effects of heavy fuel oil contamination on the growth and the development of Salicornia fragilis Ball and Tutin, a salt-marsh edible species, were studied as greenhouse experiments. Phytotoxicity assessments and PAH shoots assays were followed to measure the impact of petroleum on plant development. Syptoms like chlorosis, yellowing, growth reduction and perturbations in developmental parameters were visually observed. In this study, shoot coating appeared to be less than through soil and the investigators observed more marked effects on plants as an indication of the degree of pollution. However, a significant bioaccumulation of PAHs in shoot tissues was also found, even at very low levels of contamination. These highly related to the conditions of exposure to oil. This strong relationship between the PAH contents of Salicornia plants and growth reduction suggest a chemical toxicity of fuel oil. The investigators concluded that the type and degree of fuel oil contamination are two important factors in controlling the impact of fuel oil on S. fragilis.

In 1986, more than 8 million litres of crude oil spilled into a complex region of mangroves, sea-grasses and coral reefs, just east of the Caribbean entrance to the Panama Canal. Intertidal mangroves, sea grasses, algae and associated invertebrates were covered by oil and died soon after. Investigators reported that seedlings of red mangrove, Rhizophoram angle, which were transplanted to heavily oiled sites, did not produce new leaves. This contrasted the transplants at an unoiled site. Entire beds of intertidal sea-grass Thalassiat estudinum were killed on some heavily oiled reef flats, as shown by the abundant oil-covered dead leaves washed ashore, as well as the dead, but intact, root-rhizome mats. In contrast, the subtidal Thalassia survived everywhere after the spill. Having said this, in the heavily oiled areas, the leaves of the subtidal Thalassia became brown and heavily fouled by algae for several months.

Another example of the oil spill effect on mangrove plants is December, 2000. In this case, 500 mangrove saplings in a 6.34 ha reserve in Hong Kong were subjected to a smuggled fuel oil spill. More than 80% died in root-zone sediments containing 60 - 80 ug/g total petroleum hydrocarbons (TPH). However, one year later, the injured survivors had recovered and re-grown, with root-zone sediment TPH concentrations approaching urban 'background' values of 40 ug/g TPH.

Impact of Oil Spills on Cyanobacteria and other Microorganisms

There are numerous factors that determine the microbial response to marine oil spills. These include the oil composition and degree of weathering, as well as the environmental conditions, particularly temperature and nutrient concentrations. When crude oil is introduced into seawater, the microbial community changes and consists of multiple co-existing species. These can be explained by resource sharing. The diverse array of hydrocarbons present in crude oil requires resource partitioning by microbial populations, as well as microbial modification of oil components and the surrounding environment, will lead to temporal succession. The reviewers observed a network of direct and indirect interactions within and between species, even when just one type of hydrocarbon is present. They also provided a schematic illustration of some of the interactions involved in hydrocarbon biodegradation. Elements of these interactions were stated in several studies reviewed by them.

Multiple co-existing species and/or temporal succession are typically observed when Alcanivorax spp., which degrade straight-chain and branched alkane, are increased and found abundantly. This is followed by Cycloclasticus spp., which degrade PAHs. Other genera of obligate alkane degraders, such as Oleibacter sp. and Oceanospirillales sp., have been abundantly detected in other oil-rich marine environments.

At the molecular level, studies by the microarray analysis of 16S rRNA genes have revealed that Gamma-proteobacteria were enriched in marine environments that are historically exposed to frequent, episodic, natural 'oil spills'. This analysis also revealed that marine water exposed to oil spills enriches indigenous oil degrading bacteria, where members of the order Oceanospirillales comprised more than 90% of the bacterial community, compared to 5% of the uncontaminated sample. Experiments on the incubation and mixing of marine sediment with phenanthrene and bromodeoxyuridine (BDU), followed by the analysis of BDU-labelled DNA, revealed diverse groups of PAH degraders belonging to the genera Shewanella, Exiguobacterium, Methylomonas, Pseudomonas and Bacteroides.

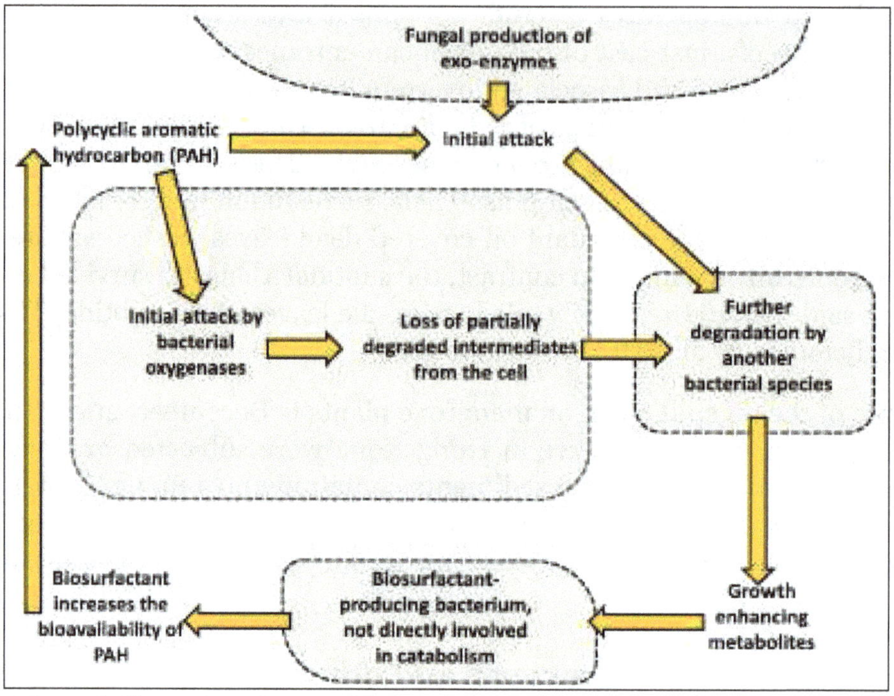

Figure: Schematic illustration of some of the interactions involved in hydrocarbon biodegradation.

It also revealed Gammaproteobacteria and Deltaproteobacteria, which were not closely related to the cultivated organisms. Following the 1991 Gulf War environmental disaster, an extensive formation of cyanobacterial mats was observed to colonize most of the oil polluted shores. Although most of the intertidal cyanobacterial mats were severely affected by the oil spill, these organisms were the first to re-colonize the destroyed habitats. This initial massive growth of cyanobacteria, especially on sites where they did not occur before the oil spill, indicated the preference of the cyano–bacterial mats to the absence of bioturbation (i.e., destabilizing the sediment surface caused by crabs and polychaetes) of the sediment for their growth. Cyanobacteria do not usually occur where bioturbation has been carried out and - together with the grazing pressure by benthic animals - prevents the establishment of cyanobacterial mats. The extensive growth of cyano–bacteria following the oil pollution of the shores can be explained by destroying most of the crab colonies in the mudflats and immediately stopping the bioturbation process, as well as grazing by gastropods. This removes the uppermost part of the oiled sediment. The second is the resettlement of burrowing macrofauna like crabs and benthic animals, such as gastropods, which out-compete the cyanobacteria again. The third is further extensive growth of cyanobacteria building thick laminated mats. These layers completely seal the surface and hence, produce an anaerobic zone which inhibits oil degradation. As long as such a bloom of cyanobacteria exists, microbial oil degradation will be prevented. They will also prevent any resettlement by macro-fauna.

Economic Impact of Oil Spills in the Gulf

During the 1930s and 1940s, the discovery of oil in the Gulf led to a massive increase in shipping. This discovery is principally responsible for the huge economic wealth and strategic importance associated with the region today. Thus, the socio-economic development of the Gulf region is highly dependent on its marine environmental quality.

In 1991, the second Gulf War led to the largest oil spill in human history. Around 6 million barrels of oil were discharged into the Gulf. Moreover, 770 km of coastline from southern Kuwait to Abu Ali Island (Saudi Arabia) were smothered with oil and tar, erasing most of the local plant and animal communities. Many of these communities were internationally significant. For example, a number of bird species, green and hawksbill turtles and dugongs are endemic to the region. Such species were harmed and disturbed.

Considering that approximately 49% of the world's oil production comes from the Gulf States and passes through the Gulf, its liability to pollution is about 48 times that of any other similar area on earth. Hence, the Gulf is possibly the most chronically oil-polluted marine area in the world, even before the war.

Due to the different anthropogenic activities relating to oil spills, in addition to the natural environmental stresses of the Gulf, i.e., enclosed and shallow nature, a number of socioeconomic impacts are predicted:

- Fisheries as a multi-million dollar industry and the artisanal fisheries as a resource of great social significance are threatened. This is because oil spills are harmful to coral reefs, mangrove areas and sea-grass. These organisms provide a support and grounds for a number of commercially significant fish and shrimp species.
- Desalination plants that provide most of the population's freshwater supply for the Gulf region are threatened.
- Disruption to the fishing industry and a reduction in scuba diving tourism. Consequently, people who have a career in this industry will lose their jobs, which they have had for a long time.
- Disruption of the coastal and marine environments which are internationally significant for a number of bird species, green and hawksbill turtles and dugongs. These are endemic to the region.

Marine Biodiversity Conservation

'Biodiversity' is one of the most common keywords used in environmental sciences, spanning from research to management, nature conservation, and consultancy. Despite this, our understanding of the underlying concepts varies greatly, between and within disciplines as well as among the scientists themselves. Biodiversity can refer to descriptions or assessments of the status and condition of all or selected groups of organisms, from the genetic variability, to the species, populations, communities, and ecosystems. However, a concept of biodiversity also must encompass understanding the interactions and functions on all levels from individuals up to the whole ecosystem, including changes related to natural and anthropogenic environmental pressures. While biodiversity as such is an abstract and relative concept rooted in the spatial domain, it is central to most international, European, and national governance initiatives aimed at protecting the marine environment. These rely on status assessments of biodiversity which typically require numerical targets and specific reference values, to allow comparison in space and/or time, often in association with some external structuring factors such as physical and biogeochemical conditions.

The term "biodiversity", first used almost three decades ago as a derivative of "biological diversity" today is one of the most often cited terms in both ecological research and environmental management and conservation. However, its precise definition and our understanding of the concept vary widely both between and within disciplines. Biodiversity is recognized to encompass ". the variability among living organisms from all sources including, inter alia, terrestrial, marine, and other aquatic ecosystems and the ecological complexes of which they are part; this includes diversity within species, between species and of ecosystems.". The elements of biodiversity are fundamental properties of an ecosystem, and, in the marine realm, these encompass all life forms, including the environments they inhabit, and at scales from genes and species to ecosystems. Biodiversity can be viewed as an abstract aggregated property of those ecosystem components and can relate to the structure or function of the community where structure relates to the system at one time whereas functioning relates to rate processes. The structural aspect is represented by the various marine life-forms, ranging from the smallest prokaryote to the largest mammal, and inhabiting some of the most extreme environments. These species exhibit a diversity that probably exceeds that found in terrestrial environments. The functional aspect is represented by the relationships among and between these marine organisms and the environments they inhabit, and is defined in terms of rates of ecological processes; most notably they include physiological processes, predator-prey relationships, trophic webs, competition, and resource partitioning. These functions vary on both temporal and spatial scales, and include some of the most important ecosystem services, including oxygen provisioning, CO_2 sequestration, and re-mineralization of nutrients. Both structural and functional elements contributing to biodiversity play a fundamental role in maintaining and defining healthy marine systems.

In essence, the marine ecosystem is comprised of three interlinked processes. Firstly, the physico-chemical system creates a set of fundamental niches (most often the water column and substratum) which then are colonized by organisms according to their environmental tolerances—these may be termed environment-biology relationships. Secondly, the organisms interact with each other in, for example, predator-prey interactions, competition, recruitment, feeding, and mutualism—these are biology-biology relationships. Thirdly, the resulting ecology has the ability to complete the cycle with feedback loops and modify the physico-chemical system through bioturbation, space or material removal or change, bio-engineering, for example; these may be termed biology-environment relationships. Superimposed on these three systems are anthropogenic influences which then perturb the systems.

Human activities produce a range of pressures on marine systems, some of which may lead to irreversible changes. This may have immediate consequences for patterns of biodiversity and consequently for the critical ecosystem services they provide. Those ecosystem services can be grouped into provisioning, regulating, supporting and cultural ones which, after adding human complementary assets, in turn lead to societal benefits.

In this context, the European Marine Strategy Framework Directive (MSFD) requires Member States to achieve Good Environmental Status (GES). The directive comprises 11 qualitative descriptors of GES, of which biological diversity is the first, but most if not all of the others can be considered to refer to some part of biodiversity in its broad sense, assuming we also consider habitats and their condition as being within the term; indeed it can be assumed that if the biodiversity descriptor has been satisfied then by definition all others are satisfactory and vice versa. In order to know whether the goal of GES has been achieved, an assessment needs to be performed that measures the current environmental status, hence this involves quantifying the abstract ecosystem feature biodiversity. For this, the European Commission has defined a number of GES criteria and indicators that represent and quantify various aspects of environmental status and biodiversity. The available indicators in Europe, for the MSFD implementation, have been recently collated, and a method to select the most adequate has been proposed. Then, some of them have been used in assessing the environmental status across regional seas.

It is axiomatic that one cannot manage a system unless it can be measured and those measures require to be SMART (Specific, Measurable, Achievable, Realistic, and Time-bounded) otherwise it is not possible to determine whether management has achieved the desired result. Hence the importance of quantitative indicators but these must be comparatively simple if they are to be operational, although many of these overlap, and such redundancies can compromise the efficiency and accuracy of assessments. The recent trend toward using long lists of indicators for an integrative assessment increases the risk of such overlaps.

Given the inherent complexity of biodiversity and the services which the ecosystems provide as a consequence of their biodiversity, it is imperative to depict these into one or more simple conceptual models. There are many ways to view marine systems, depending on the questions asked, the management goals set and typically, as with any complex system, disaggregating the various levels of complexity allows us to better understand each of the components and their major interactions. Consequently, an assessment of biodiversity used to answer a specific question will benefit from a set of conceptual models which together represent the various aspects of biodiversity. Together, these models provide a multi-faceted view of biodiversity and help users to identify the necessary elements to include in an environmental assessment by focusing on the aspects of biodiversity most relevant to the specific question and goal.

A common conceptual framework on marine biodiversity is presented here to facilitate integrative assessment of environmental status and implementation of the relevant legislation. We present a context-driven, multi-faceted view on biodiversity that will enable selection of the appropriate assessment elements and indicators. The framework is required to implement and further develop policies and practice to maintain biodiversity in the context of the sustainable management of human activities.

Conceptual Views of Biodiversity

Marine biodiversity is an aggregation of highly inter-connected ecosystem components or features, encompassing all levels of biological organization from genes, species, populations to ecosystems, with the diversity of each level having structural and functional attributes table. Further, marine biodiversity, or any of its components, can be assessed at various temporal or spatial scales. A conceptual model of marine biodiversity and its interpretation therefore depends on the questions being asked, which of the different components are emphasized, and the information and understanding available, especially of the connectivity and feedbacks in the system. By definition, this involves the implicit understanding that the components are all part of a larger and inter-linked system, where changes in one element inevitably will produce knock-on effects elsewhere. These may be regarded as bottom-up processes, causing change from the cell to the ecosystem and from the physicochemical system to the landscape ("seascape") system. Similarly, they can be regarded as the responses in a top-down system focusing on the upper level (seascape and ecosystem) which is often the end-point of marine management and the focus of the current review. Accordingly, this does not specifically address genetic, molecular, physiological, biochemical, population, and size-biomass-spectrum aspects of biodiversity, as these are both intrinsic and implicit aspects within the concept of biodiversity, whichever viewpoint is emphasized. We thus specifically cover only the upper levels, but retain the understanding of the multi-level complexity within these.

Table: Structural and functional biodiversity examples across levels of biological organization.

Level of biological organization/compositional level	Structural diversity	Functional diversity
Genes-molecular	Genetic structure, gene pool; molecular and biochemical structure	Genetic variability over time, gene pool modification; biochemical changes in space and time.
Species-individual	Morphological variability, size-biomss spectra	Physiological variability; environmental tolerance change; growth variability.
Species-population	Polulation structure, recruitment size, biomass variability	Population dynamics, production and productivity change; intra-specific relationship changes.
Community	Community composition	Inter-specific relationship changes; organism-habitat variability; intra-habitat competition; food-web interaction.
Ecosystem	Ecosystem structure	Ecosystem processes, predator-prey relationship changes inter-habitat competition.
Landscape type	Habitat structure; seascape mosaic	Physical-biota interaction variability in space and time; changes to seascape mosaic in space and time.

Hence modeling such a complex system with a view to marine management requires (i) pragmatic simplifications through disaggregation of the elements into various conceptual viewpoints, followed by (ii) a context-driven re-aggregation of the necessary components. We here provide three illustrative examples of such conceptual upper-level views on marine biodiversity, where the information retrieved is restricted to that relevant to the main focus, or viewpoint. The first focuses on structural aspects using a classical taxonomic approach to biodiversity (structural taxonomic biodiversity). The second focuses on the functional aspects of biodiversity (functional ecosystem biodiversity), and the third illustrates food-webs as one of the most used types of a combined view on both structural and functional aspects of biodiversity (food-web biodiversity). These examples only capture parts of the full complexity of biodiversity but are the most commonly found in specific user-driven contexts.

Figure: Schematic illustration of a pragmatic simplification of marine biodiversity, where a restricted extent of information is selected depending on the relevant viewpoints and questions asked.

Structural Taxonomic Biodiversity

Quantitative taxonomic data sets are a useful tool in environmental assessments, with typical indicators being species (taxon) richness, and population abundance and biomass within a place, between areas or over time. This is especially important in nature conservation planning, notably because habitat destruction is a major driver of species extinctions, particularly those with narrow distribution ranges, such that adequate knowledge of the structural taxonomic biodiversity of a particular area will help to preserve its endemic species. A taxonomic inventory and the associated habitats and their changes in space and time then becomes central to environmental impact assessments, studies of marine protected areas and the compliance with marine diversity and ecosystem health governance instruments such as the EC Habitats Directive.

The EU MSFD addresses biodiversity components within two main categories: (i) main species groups, and (ii) habitats and their associated communities (habitat diversity and mosaics). The main species-level groups include mammals, birds, fish, cephalopods, and reptiles. Within the marine habitats, water-column communities comprise pelagic microbes, phyto- and zooplankton, whereas seafloor communities encompass benthic micro, macro- and mega- fauna as well as primary producers such as sea-grasses and macro-algae. In addition, other species such as those included under the European Union legislation or international conventions, charismatic or non-indigenous species and genetically distinct forms (varieties or subspecies) of native species may be included, depending on the particular assessment area and questions being addresses. In the MSFD, the categories for birds, fish, and mammals are further sub-divided into main functional categories, mostly based on their feeding and/or depth preferences. This, however, introduces a functional division into the otherwise purely structural view.

Table: Predominant functional and/or feeding groups within the main biodiversity components for application.

Biodiversity component	Ecotype
Birds	Offshore surface-feeding birds Offshore pelagic-feeding birds Inshore surface-feeding Inshore pelagic-feeding birds

	Intertidal benthic-feeding birds Subtidal benthic-feeding birds Ice-associated birds
Reptiles	Turtles
Mammals	Toothed Whales Baleen whales Seals Ice-associated mammals
Fish	Pelagic fish Demersal fish Elasmobranchs Deep sea fish Coastal/anadromous fish Ice-associated fish
Cephalopods	Coastal/shelf pelagic cephalopods Deep-sea pelagic cephalopods

The predominant seabed and water column habitat types can effectively be characterized in terms of a pragmatic selection of the major categories under the European Nature Information System (EUNIS) scheme. The biological communities associated with those habitats can then be addressed; thus extending the conceptual view from purely taxonomic entities to higher-level structural aggregations of taxa as part of their biotope. This structural view potentially omits the functional attributes or traits of the populations and communities associated with habitats although some of the structural attributes may be regarded as surrogates (proxies) for functional ones. For example, the benthic communities can be characterized in terms of proportional representations of different traits, feeding guilds, motility, burrowing activities etc. but these have not previously been the main focus of structural biodiversity; most methods have centered on the plethora of quantitative means of defining benthic community structure. However, recognizing and measuring functional diversity within the benthos also has become of increasing importance from a management perspective.

Table: Predominant habitat types for application in assessment of Descriptor 1.

Realm	Predominant habitat type	Relationship to EUNIS habitat classes
Seabed habitat	Littoral rock and biogenic reef Littoral sediment Shallow sublittoral rock and biogenic reef	A1 + A2.7 A2 (except A2.7) A3 + circalittoral habitats in A4, infralittoral & circalittoral biogenic reefs in A5.7
	Shallow sublittoral sediment	Habitats in A5 (except A5.6) above wavebase (from 0m down to about 50-70m depth in Atlanctic)
	Shelf sublittoral rock and biogenic reef	Deep circalottiral habitats in A4 & A5.7
	Shelf sublittoral sediment	Deep circalittoral habitats in A5 below wavebase (from about 50-70 m depth down to the shelf break Atlantic)

	Bathyal rock and biogenic reef	A6.1 + A6.6 (bathyal zone- ~ 200-1800m in Atlantic)
	Bathyal sediment	A6.2 + A6.3 + A6.4 +A6.6 (bathyal zone- ~ 200-1800m in Atlantic)
	Abyssal rock and biogenic reef	A6.1 + A6.7 (abyssal zone- ~ >1800m in Atlantic)
	Abyssal	A6.2 + A6.3 + A6.4 + A6.6 (abyssal zone- ~ >1800m in Atlantic
Pelagic habitats	Low salinity water (Baltic sea) Reduced salinity water (Black Sea) Estuarine water Coastal water Shelf water Oceanic water	EUNIS pelagic classification not structured in suitable way for purpose here.
Ice habitats	Ice associated habitats	AB

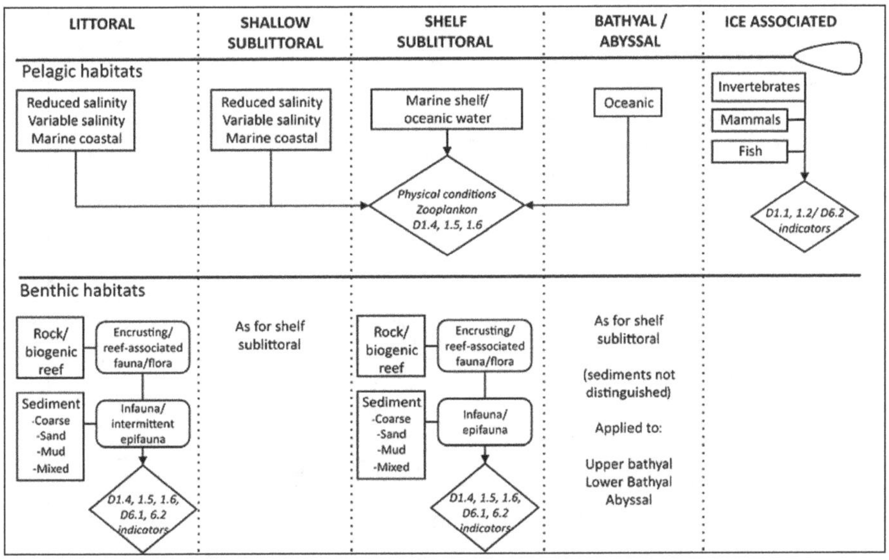

Figure: Conceptual illustration of the biodiversity components associated with pelagic and seafloor habitats.

A high biodiversity, including species richness, may enhance ecosystem processes and promote long-term stability by buffering, or insuring, against environmental fluctuations. Conversely, a loss of biodiversity may impair ecosystem functioning, and thus also the services provided. At least in the marine realm, habitat structure obviously influences the number of niches available for colonization and thus can indicate the number of types (species, traits, etc.) which can be supported within that habitat. Other community properties such as biomass and abundance are more dependent on ecological interactions such as predator-prey links and recruitment. This biodiversity-stability relation is complex as it firstly requires a clear definition of what is meant by ecosystem temporal (dynamic) stability and/or the ability to withstand change through resistance and resilience. Secondly, it requires understanding how biological diversity will enhance ecosystem stability. There is a wealth of theoretical and empirical data to support the contention that biodiversity (numbers of distinct species, but also functional diversity) enhances both ecosystem productivity and its resistance to perturbation. Habitats and species diversity are intrinsically intertwined, and baseline diversity is highly variable. For example, species diversity in seagrass meadows is greater than in adjacent non-vegetated areas, but the lack of seagrass diversity makes these habitats more vulnerable to specific perturbations such as the Wasting disease and storms. However, this is not always the case as some lower diversity ecosystems, such as estuaries, have a high resilience conferred by the high tolerances and adaptability of the component species, a feature termed environmental homeostasis.

While structural taxonomic biodiversity may enhance ecosystem stability, it is not the structural biodiversity as such that causes stability, but the individual species and their role in the ecosystem. In order to understand which species or species groups are the major players within marine ecosystems and how they relate to the functioning of the ecosystem, the understanding of biodiversity would have less emphasis on recording all the taxa, but rather on including the main species within the different functional or feeding groups. This implies a redundancy in the ecosystem, the so-called "rivet hypothesis". This also emphasizes the need for a functional view of biodiversity.

Functional Ecosystem Biodiversity

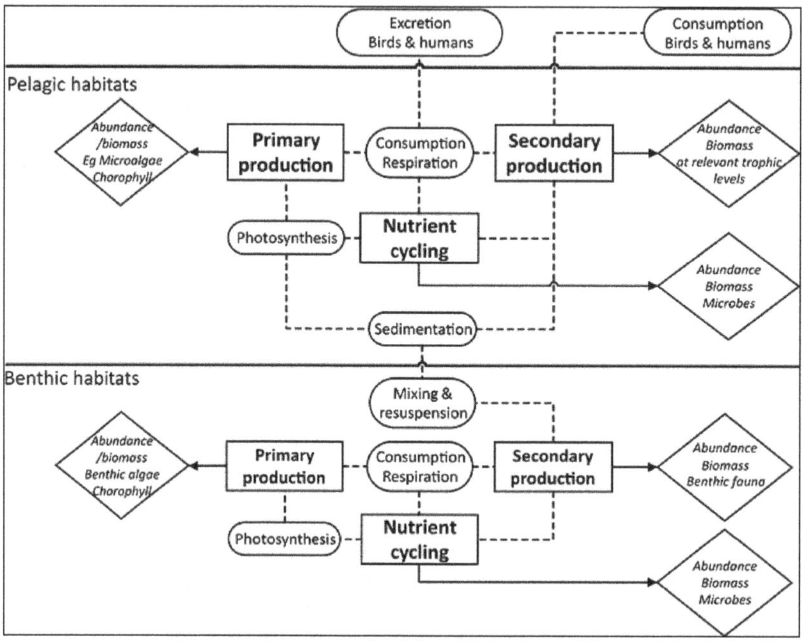

Figure: Conceptual illustration of the major functions within marine ecosystems, as a basis for structuring ecosystem-orientated biodiversity assessments.

By interpreting biodiversity from an ecosystem (top-down) entry point, the focus shifts from structural to functional aspects. In order to construct a simple-to-use view, it is necessary to distinguish between the terms functions and processes of which there are three main categories of ecosystem *functions*: (i) *Primary production*; (ii) *Secondary production* (spanning from the herbivorous primary consumers to the top predators), and (iii) *Nutrient cycling*. Each of these major functions is carried out through many inter-linked *processes*, such as photosynthesis, particle flux (sedimentation, mixing, and re-suspension) and consumption/respiration. Export of energy from the marine system to humans and birds through selective biomass extraction also is considered a process as is the re-introduction of nutrients through effluents/run-off and guano.

Documenting the biodiversity status of these three major ecosystem functions/processes, through which they are carried out, requires measurable parameters and indicators (diamond-shaped boxes in Figure). Most of the indicators currently, or potentially, used in environmental assessment are regarded as surrogates (proxies) of the three main ecosystem functions, but the extent to which these reflect the processes is variable, and often just reflect structural elements of the ecosystem. Measuring the abundance and/or biomass of microalgae, the content or concentration of chlorophyll or various proxies such as fluorescence is commonly used to represent the amount of primary producers in the system, even if these indicators do not always directly measure photosynthesis. Similarly, for nutrient cycling, appropriate indicators may include the abundance or biomass of microbes or the conservative or otherwise behavior of the different nutrient forms, but this may not give sufficient knowledge of microbial activity. Secondary production, on the other hand, is more tangible, and there exist many indicators that are proxies for quantifying the distribution, population dynamics, abundance, and condition of the various categories of organisms, both in terms of functional traits and population and taxonomic composition. Measuring the processes directly is somewhat more challenging because it often involves experimental approaches (for example respiration measurements), or long-term passive sampling (for example sediment traps) or repeated time-series of population dynamics, Allen-curves and biomass changes to allow production and productivity to be estimated, and these can be particularly time-consuming, expensive and not least of all, highly variable from daily, seasonal to annual scales.

A unified approach to a biodiversity assessment with a functional ecosystem focus would therefore start by identifying indicators for the three main functions. Most assessment programmes will not include these functions, but their existence should at least be acknowledged. From there, the key processes and taxa within each of the major functions will be identified, first in general terms, and then in detail, specific to the assessment area in question. Furthermore, it is argued that there is an increasing emphasis in marine management, from the structural ecological approach in the EU Water Framework and Habitats Directives, to the more functional approach in the MSFD.

Food-Web Biodiversity

The food-web functional view employs the three main ecosystem functions (primary production, secondary production and nutrient cycling) thus encompassing a range of processes. The three ecosystem functions are carried out by various combinations of the structural components of biodiversity. Primary producers in the form of microorganisms, micro- and macro-algae as well as macrophytes (e.g., seagrasses), and including both photo- and chemosynthesis, exist in both the pelagic and benthic realms. Through the microbial loop and remineralization, microbes are responsible for the key function of nutrient cycling and make carbon available to the system. The primary herbivorous grazers such as copepods form the link between primary production and the rest of the food-web, although these also are transported out of the strictly marine system through harvesting by seabirds and humans, as a source of omega-3 oil.

Thus, functional indicators of nutrient cycling can operate on microbes, primary production and secondary production to zooplankton, benthos and progressively higher-order predators. The processes typically are explored using more field-experimental, research-orientated indicators although the parameters or organisms to be measured within the three ecosystem functions depends on the biodiversity characteristics of the assessment area and the management questions being addressed.

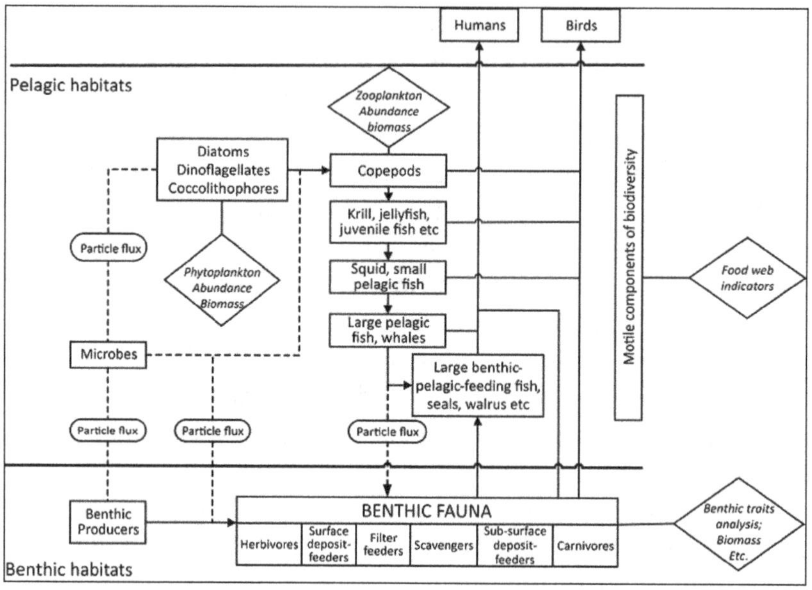

Figure: Conceptual illustration of a generic marine food-web.

In essence, a generalized food-web assessment requires indicators to cover all the major energy flow pathways throughout the system. Indicator selection would conceivably start at the producer level, such as abundance and biomass of phytoplankton and benthic algae, and also the basal zooplankton consumers. Indicators for motile components within the pelagic habitat would cover smaller components to top predators, assessed in categories appropriate to the survey area, but essentially covering, for example: (i) krill, gelatinous plankton, and juvenile fish, (ii) squid and small pelagic fish, (iii) large pelagic-feeding fish, reptiles, and mammals such as seals and finally (iv) large benthic feeding fish and mammals such as walrus and seals. The benthic secondary producing component can be seen in terms of functional groups, from herbivores (such as grazers), carnivores which actively seek prey and scavengers which consume both living and dead remains, to surface deposit feeders which consume material deposited from the planktonic realm, and filter-feeders that operate at the sediment-water interface, feeding on both settling particles as well as re-suspended matter, the latter produced either through biological pumps or strong bottom currents.

Implications for Biodiversity Assessments

Different management questions require different starting-points for selection of measurement parameters and indicators for biodiversity assessments.

Table: Examples of common managerial questions and the appropriate conceptual viewpoints, as starting-points for indicator selection for biodiversity assessments.

Managerial Questions	Conceptual viewpoints	Examples of indicators/methods	Informative value	Potential gaps
Conservation; Maximizing biodiversity.	Structural taxonomic biodiversity.	Species abundance, richness, diversity, Physical sampling and/or visual methods.	Informs of range of species present; useful as reference conditions.	Detailed observations made at large scale may not always be correctly upscaled to represent a wider area.

Eutrophication/Hypoxia.	Functional ecosystem biodiversity, structural taxonomic biodiversity.	Productivity, harmful algal blooms, seafloor species abundance, richness, diversity, indicator, taxa, substrate condition, sediment profile analyses, physical analyses of substrate (O_2 etc).	Informs of degradation status of both the habitat and the faunal communities.	Assessments shall include monitoring of water column quality, i.e., nutrient levels and phytoplankton.
Monitoring of seafloor condition/ disturbance (local scale).	Structural taxonomic biodiversity.	Species abundance, richness, diversity, indicator taxa, sediment profile analyses, physical analyses. Physical sampling and/or visual methods.	Physical sampling gives rise to quantative indicators of seafloor biodiversity and disturbance. Visual methods give a broader overview of conditions and visible disturbance (e.g., smothering or abrasion).	Visual and physical sampling can cover only a relatively limited spatial area (appropriate for localized point-source disturbance). Less informative for more spatially extensive, but less locally intensive disturbances.
Monitoring of water column quality.	Functional ecosystem biodiversity.	Abundance/biomass e.g., of microalgae, chlorophyll. Use of physical sampling and/ or remote or in-situ sensors, biomarkers, areal or satellite monitoring.	Information on water quality (parameters as relevant), early warning system of change, biological effects monitoring.	Physical sampling or infrequent remote measurements will not capture short-term fluctuations, but in-situ sensors will do so. Organisms for bio-makers integrate conditions over time.
Protection of coral structures.	Structural taxonomic biodiversity.	Species abundance, richness, diversity. Reliance on visual and acoustic methods; no physical sampling.	Acoustic methods can localize coral structures over larger areas, and visual methods used to verify potential finds.	Visual methods allow identification of corals and larger epifauna (and fish), but will underestimate abundance and diversity of burrowing or smaller organisms utilizing the coral habitat.
Sustainable human activities (broad-scale).	Functional ecosystem biodiversity, food-web biodiversity.	Abundance and/or biomass of primary producers (incl. microbes). Productivity of key species or trophic groups, proportion of selected species.	Holistic assessment of biodiversity at a broad, ecosystem scale. Useful also for determining large scale impacts of local disturbances.	This topic is extensively so likely no monitoring program will cover all of these issues. A more detailed question-driven selection of indicators will be required.

Structural Biodiversity Assessment

The structural view on biodiversity is typically used when nature conservation is the primary focus in preserving all (or at least those designated as being important) biotic components of a given ecosystem together with its characteristic abiotic features. For example, the EC Habitats Directive requires assessing the biodiversity status, especially for the conservation features for which an area was designated, by using the appropriate taxonomic and habitat quality indicators. This either ignores the functional relationships within the ecosystem or makes the assumption that the structural elements are proxies for functioning. This can have implications for the management of such conservation areas since it may require manipulating the habitats and living conditions of certain species or communities when the assessment reveals a less favorable biodiversity status. In this case, ecoengineering may be required both to recreate and restore suitable eco-hydrological functioning (Type A ecoengineering) or to use the restocking or replanting to recreate populations (Type B ecoengineering). As an example, reef restoration is a measure to re-establish reef systems in places where these might have been damaged or lost. This requires the current habitat to be altered (e.g., from soft bottom to hard bottom) so it can support and promote the establishment of a new reef community. This structural change will be reflected in later biodiversity assessments and possibly document the increased biodiversity status. However, if the focus is on a structural view of biodiversity, it might not result in successful functioning and so this kind of biodiversity assessment will not be a holistic one. Hence, the context-driven approach maximizes taxonomical biodiversity but not necessarily ecosystem functioning. Although it can be assumed that biodiversity and ecosystem functioning relationships (BEF) will ensure that higher taxonomical biodiversity also produces higher ecosystem stability (in terms of resistance and resilience), there is insufficient evidence to support this assumption.

Ecosystem Assessments

Most management policies and assessments world-wide aim for some kind of ecosystem approach. The MSFD advocates an ecosystem-based approach, and many assessment and monitoring schemes exist aiming to integrate ecosystem functions and their values and services. However, as with the term biodiversity, the distinctions and uses of the terms Ecosystem Approach and Ecosystem-based management are far from consistent. An Ecosystem-based management strategy acknowledges the complexity of ecosystems and in particular: (i) the need to take into account both the structural aspects (e.g., life-forms present) and the interactions among organisms (especially inter-species relations) within ecological systems, (ii) the essence of connectivity between and within communities, ecosystems, habitats and biotopes, and (iii) that humans are a part of ecosystems thereby integrating human societies within biodiversity management. This approach encompasses the structural and functional aspects of an ecosystem (its "emergent properties") as well as, at a smaller scale, the role of given subsystems or components from this ecosystem.

To that end, ecosystem assessments tend to employ at least two views on biodiversity: The structural taxonomic and the functional ecosystem biodiversity. Both are used, or at least require to be used, in one single assessment, but require the need to keep overlaps minimal and to properly interpret the results when measures are to be taken on the basis of the assessment results. This, in turn, requires the need to interpret the resulting ecosystem status in both structural and functional ways so that managers can balance the different needs when planning management measures. As an example, proposed an ecosystem health assessment (or monitoring) programme consisting of four elements associated to the typical management cycle: (i) an analysis of main processes and structural characteristics of an ecosystem; (ii) an identification of known or potential stressors; (iii) the development of hypotheses about how those stressors may affect each part of the ecosystem, and (iv) the identification of measures of environmental quality and ecosystem health to test hypotheses. This encompasses and quantifies, from the socio-ecological system, the ecosystem services, and societal benefits approach. This approach has led to an extensive series of marine assessment systems which can include both the ecological health and societal well-being, for example the global Ocean Health Index (OHI).

In general, starting from the conceptual view of functional biodiversity, the clear distinction between ecosystem function and process (e.g., as proposed above) must be retained throughout the assessment and its interpretation when the terms are used to derive management actions from the indicators used to assess functions and processes. However, there is a notable lack of agreement throughout the literature regarding the terms "function" and "processes" when applied to ecosystems and their assessment; indeed the terms may be synonymous in that by definition a function is a rate process. In our functional ecosystem model, the three ecosystem functions (primary production, secondary production and nutrient cycling) together comprise holistic ecosystem functioning. These ecosystem functions are the sum of the physical, chemical and biological processes that transform and translocate energy and materials in ecosystems.

Functions, and thus inherently also the processes by which they are carried out, are central to the "ecosystem services" which the marine environment provides for its own sustainability and human benefits. As indicated above, successful structure and functioning of the physico-chemical and ecological systems can produce intermediate and final ecosystem services:

- Provisioning,
- Regulating,
- Supporting (or habitat),
- Culture and heritage.

Complementary human assets are then required to extract societal benefits from such services. Strong et al. listed five categories of "ecosystem functions," which also refer to processes:

- Production of biomass,
- Organic matter transformation (non-living),
- Ecosystem metabolism,
- Nutrient cycling,
- Physical environment modification, for which they analyzed biodiversity.

Thus, there are many ways to refer to the functions and processes occurring within marine ecosystems, and in turn the services and societal benefits which they provide. Focusing our conceptual understanding of biodiversity from a functional ecosystem viewpoint on three main functions, driven by a range of processes, gives clarity about the logical basis for both selection of assessment parameters and interpretation of results. We recognize that the functions themselves are assessed by measuring some proxy of the processes, such as various qualities and attributes of the organisms which carry out those processes. With this understanding, we can select the indicators which represent the sections of the system which best address the questions asked, and at the same time retain an awareness of the information gaps which require us to extrapolate information from other measurements and to make appropriate inferences for ecosystem-scale assessments.

Food-Web Assessments

The conceptual view outlined in Figure provides the basis of a holistic food-web assessment. Typically, such assessments operate with a restricted set of parameters relating to predator-prey interactions, with a focus on abundance and population structure of commercially harvested species, and often also their main prey items. For example, the MSFD Descriptor 4 (trophic relations) adopted a pragmatic conceptual simplification in approach. Two key attributes for food-webs were specified within the MSFD as: (i) energy flow in food-webs, i.e., from primary to secondary production, and (ii) structure of food-webs i.e., size and abundance of predators/prey argued that three main properties of food-webs can be considered within the MSFD context: Structure, functioning and dynamics, with emphasis on the latter two and "the general principles that relate these three properties." The MSFD Descriptor 4 indicators for food-webs, such as the reproductive success of dominant piscivorous seabirds, are very much process-based and designed to capture responses to the multiple anthropogenic pressures that can affect food-webs, the main one being selective extraction of biomass (e.g., fishing).

The structuring influence of large predators on ecosystem stability, and the potential for human impacts thereon, can be illustrated, for example, by overfishing of the Atlantic cod, *Gadus morhua* which caused a notable increase in alpha and beta diversity of the remaining fish communities. These became more variable during periods where the cod no longer dominated the system. This is an example of the difficulties a biodiversity concept will face when it becomes more complex. The overall assessment result will no longer be able to reflect both the structural and functional changes individually. The representability of an assessment of food-web status thus depends much on the indicators chosen and whether they are capable of capturing the "health" of the ecosystem, in terms of deviation from reference or target conditions (assuming these are in fact known and/or defined). The concept of ecosystem health is integral to management questions based on the overall assessment which thus encompasses an assessment of both biological diversity and the delivery of ecosystem services and societal benefits.

Where the aim of assessment is toward sustainable management, such as in the MSFD, or marine conservation, the selected food-web measurement parameters and indicators must focus on detecting the impacts of anthropogenic pressure. However, for a programme to understand the overall predator-prey structure in a system, all levels of interactions should be included into the underlying view on the biodiversity as the basis of the assessment. As with all aspects of biodiversity, changes in abiotic conditions such as climatic ones will also impact food-webs and create moving baselines against which changes in biodiversity are judged. They are drivers for changes in species distributions, recruitment success and competition and so food-web indicators should operate at the species level (e.g., population indicators) but also at the ecosystem level when considering overall energy flow through the system.

The main practical challenge in finding fit-for-purpose food-web indicators is the variability in pressure-impact relationships on their structure and functioning. An example on how to reach a more simplified generalization is the "fishing down the food-web" rule. It proposes that fishing a food-web would first target larger and higher trophic level carnivorous fish and then progressively those at lower trophic levels, theoretically shortening food-webs. Thus, the mean trophic levels of consumers would be lower in an overfished food web, relative to an undisturbed one. An indicator reflecting the mean trophic level will adequately captures this aspect but other indicators will be needed when the aim of the assessment is not only to maintain sustainable fisheries, but also to preserve structural biodiversity. The corresponding conceptual view of biodiversity should be the basis of such preservation aims by including the relevant structural elements into the food-web but also assuming that such structural indicators are indeed proxies for successful functioning.

This concept of marine biodiversity is based on three conceptual views of the upper-level aspects of biodiversity (structural taxonomic, functional ecosystem-based, and food-web biodiversity). They form the basis for constructing different biodiversity assessment types, depending on the context in which the assessment is used. The conceptual views serve as simplified common denominators from which can be developed a dialogue between both scientists and managers, balancing the needs for a sound scientific foundation and the pragmatic requirements for practical management of marine systems. The examples presented in this conceptual framework and the consequences for the assessment of biodiversity lead to three conclusions which improve the applicability and value of biodiversity status assessments and management.

Firstly, marine ecosystems are considered from different perspectives given the absence of a common and single understanding of what is marine biodiversity. The way in which we view this abstract biodiversity depends on various variables where this complexity can be simplified when focusing on the structural and functional elements of biodiversity that are important for the management question to be answered. This is best done using a carefully defined set of biodiversity elements to be assessed, knowing which elements to ignore and why and what consequences this has for the subsequent biodiversity assessment. This approach will allow for a context-driven assessment, where the meaning of the assessment result is pre-defined and derived from our applied understanding of biodiversity. The result does not need a special interpretation and is tied directly to the question we want to answer.

Secondly, we use the perspectives to construct a "management-friendly" assessment: A biodiversity status of "good" or "not good" needs a context for interpretation. This context is given by the specific conceptual view. Together, this will provide information on what is the biodiversity status and how it can be improved by managing identified problems. Only an assessment that can explain the resulting biodiversity status and give insights into how the situation can be changed following management measures is useful for management. It is the conceptual view that leads to insights and measures to be applied by management thus emphasizing the need for knowledge on the biodiversity status and where and how it requires to be improved if it is considered to be degraded.

Thirdly, be aware of the limits and degree of quantification of the assessment: Since we know what has been omitted from our conceptual view, we also know what management cannot expect to achieve. Similarly, the success of management measures and their efficacy can only be determined by quantifying the conceptual approach. A primarily structural taxonomic view of biodiversity will not lead to an assessment that points to measures improving ecosystem functions. However, the conceptual view chosen allows us to determine the limits of our understanding of biodiversity and thus the possibilities of the management measures even before the assessment has been made. If the limits are clear and can be communicated, expectations are realistic whereas unrealistic expectations may arise from an incomplete conceptual approach or false assumptions of the links between structure and functioning.

A given conceptual view can always be expanded by including more elements and shifting the focus closer to the question asked. As one example, we can include activities which create the major pathways of human pressures, the state changes they involve in the marine system and the impacts this has on society, its welfare and well-being. Such modifications will expand our understanding of biodiversity using the influential parameters relevant for the specific purpose of the individual biodiversity assessment.

Factors Affecting Marine Biodiversity

Whereas the marine ecosystem gets our attention for different reasons, the fact remains that consistent increase in the level of pollutants has degraded water bodies across the world to levels of irreparable damage. Studies explain that 80 per cent of marine pollution originates on land, in different forms of pollutant, mostly as a result of various human activities. However plastics contamination poses the biggest threat, the list of pollutant to the marine environment remains long with sewage, pesticides, industrial chemicals and other garbage.

It is true to say that despite taking numerous steps to mitigate the effects of marine pollution, there is still a long way to go to protect and conserve our water bodies. As important as finding ways to clean our oceans and lakes, creating awareness among people towards the protection and conservation of the marine environment is also a crucial effort. And being aware of a problem means knowing the issue at the grass root level. Thus, to prevent marine pollution, one must be aware of the pollutants that pose threat to the ecosystem and the sources those originate.

Here, we have enumerated 11 main causes of marine pollution which have been a troubling marine environment for quite some time now.

Ocean Dumping

Dumping of waste materials from industries, ships and sewage plants into oceans has polluted the marine ecosystem to a great extent. The oceans have been targeted as a dumping site of sewage, chemicals, industrial waste, garbage, and other debris from the land. Only mining companies across the world dump 220 million tonnes hazardous waste directly into our water bodies every year. Similarly, it is important to note that around two-thirds of the marine lives in the world have been under threat from the chemicals we use every day, including household cleaners. Since we depend on the marine ecosystem extensively, the adverse effects of ocean dumping have not only been felt by marine life but by humans as it poses health risks.

Land Runoff

One of the major sources of ocean pollution is the waste comes from the nonpoint source, which happens as a result of runoff. Surface runoff from both agricultural land and areas carry soil and particles mixed with carbon, phosphorus, nitrogen and minerals, posing threat to the marine life in alarming scale. Crossing streams and rivers, the water filled with these toxic chemicals land in the ocean, resulting in harmful algal blooms. This kind of water pollution threatens the species of fish, turtles, and shrimp etc. and also humans through the food chain.

Dredging

In this world of ever expanding industrial activities, dredging is an important activity that enhances marine transportation and other related activities. However, dredging has been a major cause of disturbance in the marine ecosystem for many years. As dredging is to remove the deposits submerged underwater, the activity alters the pre-disposed composition of soil, leading to the destruction habitat of creatures and organisms. Similarly, dredging of contaminated materials will result in the regrouping of harmful particles and contaminate large of areas of water bodies. Though steps have been taken to mitigate the effects of dredging on the marine environment, several cases involving the destruction of underwater lives are still extensively heard about.

NOx and SOx

Nitrogen oxides (NOx) and Sulfur oxides (SOx) – the two main pollutants found in shipping emissions- has badly affected both marine environment and ozone layer in a number of ways. Both NOx and SOx are combustion products that are emitted into the environment in the form of smoke. It is estimated that in 2005, the water bodies around Europe witnessed 1.7 million tonnes of sulphur dioxide (SO_2) emissions and 2.8 million tonnes nitrogen dioxide emissions from international shipping. These type of air pollution from shipping accounts around for 50,000 premature deaths per year in Europe. However, strict rules have been formed to reduce their levels in ship's emissions. With IMO revising its standards on the sulphur content of marine fuels, the vessels passing through a Sulphur Emission Control Area (SECA) are not permitted, since 2015, to use fuels with more than 0.1% of sulphur. Similarly, the sulphur limit applicable to all marine fuels used internationally will go from 3.5% to 0.5% since 2020.

Ocean Acidification

The issue of ocean acidification is quickly becoming a threat to both marine lives and humans. Ocean acidification is the continuing decrease of seawater pH caused by the absorption of carbon dioxide (CO_2) from the atmosphere. Ocean acidification has the power to greatly endanger the lives of marine organisms and also humans who depend on fish and fish products for their daily sustenance.

Sea Water Level Rising

Global warming is alarmingly increasing seawater levels, threatening the marine ecosystem. According to reports, the annual rate of seawater rise during the past two decades has been 0.13 inches a year, which is around twice the average speed of rising over the preceding 80 years. Thus, it is high time that we educate ourselves about the causes and effects of seawater level rising to save the marine environment and biodiversity. The rise in seawater level means more wetland flooding, destructive erosion and agricultural land contamination and more importantly a serious threat to the habitat of several plants, fishes and birds.

Ozone Depleting Substances

Ozone-depleting substances such as CFCs and Halon along with other pollutants from ships are destroying the ozone layer. Ozone Depleting Substances omitted by ships across the world include Methyl Chloroform, Methyl Bromide, Bromochlorodifluoromethane and Bromotrifluoromethane etc. These man-made gases are capable of destroying ozone and in effect, these gases causing harm to the marine environment in several ways.

Waste Pollution from Ships

Apart from other pollutants such as oil and gas, the waste and garbage generated on board ships poses a significant threat to the marine ecosystem. Both solid and liquid waste in form of ballast water, grey water, food waste, dunnage and packing material, paper products and cleaning material and rags etc. pollutes the seawater and badly affects marine lives. The vessels used for various purposes- be it a container or cruise ship- contribute to this pollution in different levels.

Noise Pollution from Ships

It has been scientifically proved that the noise generated from shipping operations is harmful to marine organisms. Harmful effects of noise pollution on marine organisms include haemorrhages, changed diving pattern, migration to newer places, and damage to internal organs and an overall panic response to foreign sounds. Source of noise pollution from ships include everything from engine noise to the entertainments in cruise ships. The intensity of noise pollution is higher in marine environment since noise travels greater distances easily in water and at the same time, marine life is extremely sensitive to noise due to their heavy reliance on underwater sounds for basic life functions.

Oil Spills

The world has witnessed several oil spill disasters that have been one of the major concerns of pollution to the marine environment. Disasters such as the Exxon Valdez Oil Spill and Deepwater

Horizon etc. have resulted in the extreme pollution of the marine ecosystem, killing thousands of marine species. The oil spilt destroys the insulating ability of several marine species and also the water repellency of bird's feathers, exposing these creatures to life-threatening risks.

Plastic Pollution

It is important to mention this environment degrading agent separately for the sole fact that it has and is the reason for several environmental problems both at sea and land. The ill effect of plastic pollution is wide-ranging. The plastic pollution has a direct effect on wildlife as it- plastic bags, fishing nets and other debris-chokes tens and thousands of seabirds and sea turtles every year. The ingestion micro plastics fish and other species also pose risk to their life as well as humans.

References

- Marine-biology: byjus.com, Retrieved 17, March 2020
- Marine-organisms-and-adaptations, resources-142: sciencelearn.org.nz, Retrieved 09, May 2020
- Marine-life-definition-and-examples-2291890: thoughtco.com, Retrieved 12, January 2020
- Aquatic-Ecosystems-Marine-Types-310021416: researchgate.net, Retrieved 02, August 2020
- Marine-ecology, conservation: marinebio.org, Retrieved 18, June 2020
- MARINE-ECOLOGY-309784699: researchgate.net, Retrieved 21, February 2020
- Causes-and-effects-of-ocean-pollution: conserve-energy-future.com, Retrieved 11, July 2020
- 11-threats-to-marine-environment-you-must-know, environment: marineinsight.com, Retrieved 16, April 2020

Chapter 2

Marine Vertebrates

Marine vertebrates are the vertebrates or animals with a backbone that inhabit the marine ecosystem and environment. Marine or saltwater fishes, marine reptiles, marine mammals, seabirds and marine amphibians are classified under marine vertebrates. Each topic has been thoroughly examined and explained in this chapter to provide the reader with a better understanding of marine vertebrates.

Marine Fishes

Fish are very diverse with many adaptations to the different habitats provided by the sea and the continental waters. They can be found from freezing temperatures to the tropics, and have a wide range of life histories. The maximum diversity is found in tropical littoral areas, probably related both to the stability of the oceanographic conditions and the complexity of the habitat. Diversity is maintained through predation and competition. Most fishes have a pelagic phase as eggs and larvae that ensure dispersal and colonization of new territories. The fish range from typical k-strategists with low fecundity, parental care and late maturity to r-strategists with fast growth and high fecundity with pelagic millions of eggs. The eco-biological adaptations of these ancient vertebrates to their habitat are striking and have enabled them to colonize the more extended habitat of the planet.

Figure: Differences in shape between fish from different habitats.

Fish are the most ancient and numerous of vertebrates, exhibiting great diversity in ecology, life history patterns and morphology. Adult fish range in size by three orders of magnitude from gobies

(Aphia minuta) which mature at less than 10 mm to whale sharks (Rhiniodon typus) of 10 m or more. Fish live in waters with average temperatures ranging from below 0°C to above 30°C. Over half of the recognized vertebrate species are fish and more and more are being described every year. Currently, there are around 25,000 valid species, classified into 482 families; about 60 % of which live permanently in the sea. There are also a small number of species dwelling both in fresh and marine waters: the catadromus species, such as eels, live in fresh water but enter the sea to breed; and the anadromus species, such as salmon and sturgeons, live in the sea but enter fresh water to breed.

There are five main classes of extant fishes, one of them the Chondrichthys, excluding Chimaeriformes, are collectively known as the elasmobranchs and have cartilaginous skeletons lacking a swim bladder. 96% of extant fishes are Teleostei and have bony skeletons; many of them have a swim bladder which enables them to sustain buoyancy. Studies in temperate coastal areas have found that abundance patterns of fish are influenced by the proportional representation of different habitat types. Changes in the composition of fish assemblages can, therefore, represent changes in the representation of habitats.

The diversity of marine fish is higher in coastal and littoral waters to depths of 200 m where approximately 11,300 species are found. The diversity is highest in the IndoWest Pacific, particularly in the areas between Papua-New Guinea and Queensland, Australia. Coral reefs have the highest fish diversity on an area-specific basis and 4,500 species are known from coral reefs and associated habitats. Approximately 130 marine species like tunas have circumglobal distributions in tropical oceans. From paleontological records and fish catabolism, it has been deduced that fish (finfish and elasmobranchs) are secondary sea dwellers, having had an evolutive period in freshwater.

Fish abundance and diversity arises not only from the evolution caused by competition between predator and prey but also between species needing the same resources. Living space may be more limiting than food supply in littoral rock and marine reef communities, whereas food supply may be more limiting in the pelagic system, both in prey species and possibly in their predators too. Food may be limiting at a particular stage of the life cycle, for instance during early larval development. On marine reefs all stages are found, from fine resource-partitioning - indicating competition has occurred in the past - to overlaps with several species sharing foods - indicating little or no competition (although the foods may be used differently) - to the presence of alternative species utilizing the resources, resulting from stochastic recruitment to the living space on a patch reef. Living space concerns protection from enemies and facilities for spawning as well as for food collection. Even for pelagic fish, space may be more structured than is comprehended at present; fish may be kept in certain water strata by chemical and physical preferences and by social interaction.

Fish are part of the ecosystem entering the flux of energy at different levels of the food chain. They can be phytoplankton feeders at the second link of the chain, or they can be apical predators like big tunas or sharks. Moreover, they can be fish predators or even cannibals that in turn are also eaten by fish. Marine mammals and birds also consume fish. Their remains reaching the marine floor sustain an important population of scavengers, such as small crustaceans. Fish can have an important influence on the dynamics of other organisms through herbivory and predation. Their feeding and excretion, as well as their role as prey, can make an important contribution to the trophodynamics of the ecosystem.

Besides the ecological importance of fish as components of the ecosystem and as part of the energy transfer in the food chain, man in fisheries ranging from a small- to an industrial- scale heavily exploits fish. This chapter is centered on marine fish, their behavior, biology and life histories, the population dynamics and exploitation are briefly presented, due to the limitation of space. The population dynamics, fishery, fishery management and socio-economic aspects of fish exploitation are not included although the references include further reading on these subjects.

Fish Life

Fish have adaptations to the fluid medium in which they inhabit, such as the hydrodynamic shape of all active swimming species. Any submerged organism experiences upthrust, given by the weight of the displaced fluid medium. Thus, in order to stay at the bottom of a water column, the density of the organism has to be greater than that of water; in fact most animal tissues are denser than water. Staying in open water is thus a problem that requires adaptations to allow equilibrium between body weight and buoyancy plus hydrodynamic lift. For instance, many fish have special lipid tissues that aid buoyancy since they are less dense than water. Another adaptation is to have a swim bladder filled with gas. Some fish have reduced the density of their bones to reach a density closer to that of water. The location of the buoyancy system (the swim bladder or the fat tissue) has to be in balanced positions around the body center of gravity, otherwise it will hamper the trim of the fish, resulting in a head-up or headdown position. All these adaptations are costly, requiring energy particularly to maintain the synthesis and resorbtion of the gas. Fish larvae maintain buoyancy by their small size and density, although they may have extended fins or other adaptations to increase water lift.

Life history strategies result from natural selection for species to produce the maximum number of young surviving to maturity under the conditions imposed by their biotopes (i.e. to maximize their fitness). Natural selection would be expected to favor non-reproductive activities at the expense of reproduction only when they advance reproduction at a later stage in the life history and thereby maximize overall fitness. The result of this is that organisms under different selection pressures will have characteristic life histories.

Fish display a remarkable range of life histories, from small pelagic gobies like Aphia minuta, which mature at a few months of age and die after spawning; to rockfish Sebastes spp., which attain maturity after 10 or more years and continue to spawn every year for three decades or more. Natural selection favors the individuals that make the greatest contribution to future generations. Ideally, a fish would mature early at a large size and produce numerous and large offspring over a long reproductive life span. However, in reality, resources are limited and when individuals allocate resources to improve one's aspect of fitness; there will be a cost elsewhere. The evolution of these diverse life histories result from the different ways of combining growth, maturity, fecundity and egg size to maximize fitness, depending on physical and ancestral constrains and costs and benefits set by the environment.

Most teleosts have a bipartite life cycle, where early life is spent in the plankton, generally as a larval form which bears little resemblance to juveniles or adults. The newly hatched larvae generally have positive phototropism moving towards the shallow waters. Fish larvae are lecitotrophic, starting to feed when the yolk-sac is absorbed. The duration of the larval phase is variable, from days to months, depending on the species and the water temperature. In Antarctic fish, reproduction frequently occurs in winter with long embryonic and larvae periods, the pelagic phase reaching up to two years. Some Pacific reef-fish can postpone settlement at the bottom until a suitable place is found. The duration of the planktonic phase and the possibility of moving from one water mass to another allow the colonization of new territories or the replenishment of depleted stocks.

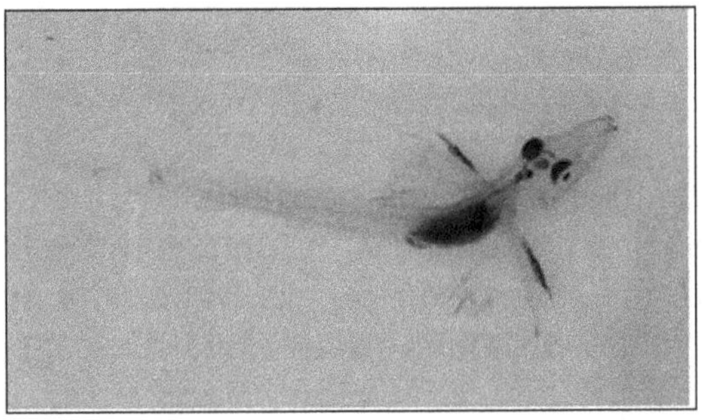

Figure: Channichtyidae larvae.

Reproduction

Fish do not reproduce at a fixed age or size, but along a trajectory of age and size that depends on demographic conditions and is determined both by genes and by the environment. The reproductive ratio is a function of body size in short-lived species, but depends on the supply of energy surplus to maintain requirements (except when feeding is very poor) in long-lived species. Many fish species, however, are very flexible in their behavior and can change allocations of resources to growth or reproduction according to environmental or social conditions.

Sexual reproduction can provide an advantage because it creates and perpetuates genetic diversity and allows the formation of novel genetic combinations that provide adaptations to ever-changing environments. Fish have separate sexes, but some can be sequential hermaphrodites. All elasmobranchs have separate sexes throughout their lives and population sex ratios are usually close to 1:1. Most temperate fishes are dioccious, but almost 50% fish families in the tropics contain hermaphroditic species. Sequential hermaphrodites may be either protogynous - in which adult females change to adult males - or protandrous - in which adult males change to adult females. The Lethrinidae, Scaridae, Labridae, Sparidae and Serranidae contain protogynous species. Protandrous hermaphroditism has been observed in Sparidae, Centropomidae and Platycephalidae. Selection should favor sex change with increasing size or age when one sex gains a relatively greater reproductive advantage with size or age. For instance, in species such as the protogynus razor fish (Xirichthys novacula) in which males defend a territory and attract females to mate, larger males have an advantage in sustaining a territory and in guarding the egg nests. In hermaphroditic species the sex ratio is biased depending on the length, hence the fishery might affect the sex ratio due to the removal of larger sizes. In the case of X. novacula, the fishery removes larger fish that are males, resulting in females turning to males at lower lengths. However, the smaller resulting males are less successful in guarding the nests and attracting females, thus producing a lower recruitment.

Sexual dimorphism in size, shape and behavior is frequent. The extreme case is the minute parasitic males of Photocorynus and Ceratias which live attached to the female body. In others there is only a difference in growth rates and longevity, like in the case of Merluccius females which get bigger and older. In elasmobranchs there is sexual dimorphism with males having claspers. The age at reproduction depends on the species and gender: most species do not mature in their first year of life and need to reach a determinate size in order to start maturing. The reduced resources available at midwater and depth have necessitated the evolution of many reproductive adaptations, such as hermaphroditism, extremes in sexual dimorphism and sex ratio adjustment.

Fecundity, Maturation and Spawning

Fecundity, defined as the number of eggs in the female prior to the next spawning period, varies very much in individuals of one species, but in general it increases as a function of the weight of the fish, hence with the cube of the length. When the number of eggs is a premium, the female is larger than the male, although in nest guarding species, the guarding sex is the larger one.

Few fish spawn only once in a lifetime; the majority spawn at repeated intervals. Reproduction is generally external, with males and females liberating the sexual products in dense clouds in the water. In elasmobranches, reproduction is internal. In some littoral fish, there is nest construction and guarding, either by the male or the female or both, depending on the species. Some species collect the eggs in pouches, like the male seahorse (Hippocampus), or in the mouth (Tilapia) and guard them until hatching. Sexual reproduction might be continuous through the year or be limited to a short period. Some species lay all their eggs at the same time, while others are serial spawners, laying several batches of eggs during the reproductive season. Maturation and spawning are synchronized through environmental parameters such as temperature or light cycles. Some are synchronized with the lunar phases: the atherinind from California Leuresthes tenuis spawns at high tide on the beach and the eggs hatch at the next tide. Lunar rhythms are also shown by Clupea harengus and Gadus capelanus.

Table: Summary of fish reproductive strategies. The r-strategists are at the top of the table, while the k-strategists are at the bottom.

Type and fecundity	Seasonality of reproduction	Examples	Type of eggs and parental care
Big bang	Once in a lifetime	Anguilla Aphia minuta	Pelagic Buried in sandy bottoms No parental care
Total spawners	Very seasonal, annual or biannual	Lates niloticus	No parental care
	Seasonal for 5- 10 years	Gadus morhua	10,000 to 10,000,000 pelagic eggs/year
Batch spawners	Broad spawning season, producing several batches of eggs	Coryphaena hippurus	100,000 eggs/batch at least 3 reproductive periods/year for 2-3 years starting at age 0
Small-brood spawners	Short annual spawning at 5-9 years of age, for 5-10 years	Harpagifer antarcticus	200-300 benthic eggs/season, parental care
	Every 1-2 years for 10-15 years	Lamna nasus	1-5 living offspring

Fecundity is also highly variable depending on the species: external reproduction is associated with lower success due to the dispersion of the gametes and the high mortality of the eggs. Some sort of egg care or incubation results in higher embryonic survival. So, depending on the reproductive behavior, the females might produce millions of eggs or only a few having typically K or R strategies. The number of eggs, their size and quality depend directly on the size of the female. It has also been found that the experience of the female increases the reproductive success. One of the effects of fishery exploitation is that females get mature at smaller sizes. Since the fishery removes the larger fish, the females can only reproduce once or twice. The combined effect of smaller, inexperienced females results in reduced stock reproductive potential, thus increasing the risk of stock collapse.

A number of fish groups produce living young. Viviparity has evolved repeatedly and independently from oviparity in many fish groups. From the 54 families of extant fishes bearing living young, there are 40 families of chondrichthyans, Latimeria and 13 teleost families with around 500 species.

There is no evidence of a phylogenetic trend that suggests a tendency to evolve towards the production of larger eggs. The partitioning of reproductive output between the size and number of offspring varies independently of body size. The general rule seems to be that pelagic spawning marine fish produce many small eggs and demersal spawners produce fewer, larger eggs. The colonization of the oceanic environment involved the production of pelagic eggs, a trend that is also found in deep-water fish which inhabit the more extended marine habitat.

Population Dynamics

Fish have a spatial distribution which is not constant in time; populations might change depending on environmental and biotic parameters, denso-dependent relationships and behavior. The density distribution of the fish in space is studied by population dynamics. Fertility (number of eggs produced), birth and death rates, life span and especially the reproductive period, are relevant parameters to determine the population abundance and composition by age groups. The environmental variability determines, in part, the population dynamics through mortality effects. The biotic population potential is expressed by:

$$Dn/dt = rN$$

where dN is the derivative of the population density increase function with respect to time t and r is the growth rate of population N. At the beginning, the population growth is exponential, becoming sigmoid with time. Due to limiting factors, the population growth decreases until an equilibrium is reached:

$$Dn/dt= Rn ((K-N)/K)$$

Where K is the asymptote of the sigmoid curve, there is the maximum number of individuals the environment can support. The denso-dependent limiting factors are space, food availability, natural mortality, and the independent ones are environmental parameters. The limiting factors express the ambient resistance against population growth and they regulate the numerical coexistence of the species constituting the community. They act on all the life phases (eggs, larvae, juvenile and adult) so that only a certain number reach the adult stage. There are K and R strategies, the former favoring the adult (growth, low fecundity, parent care and low mortality), while the R strategy tries to produce a very numerous brood, like in Gadus aeglefinus which produces 4 million eggs by spawning.

The population biomass will depend on a numerical factor, the recruitment of new fish through emigration or additions of juveniles; and a weight factor, depending on the growth of the individuals; and will decrease depending on the loss of individuals either by mortality or by emigration.

Marine Reptiles

Reptiles are the most diverse terrestrial vertebrates with about 12,000 described forms, including about 9,350 currently recognized species and about 3,000 subspecies. About 260 million years ago reptiles evolved from aquatic amphibians and by the Jurassic (150–200 myr) modern reptiles had appeared. However, only a few reptile groups re-entered the oceans, primarily sea snakes (elapids related to cobras and kraits), and sea turtles. All major reptile groups, i.e. the snakes, lizards, turtles, and crocodiles, have at least a few members that enter marine habitats even though they may have never completely adapted to a life in the open sea. Here we give an overview of those reptiles that are found exclusively or at least occasionally in the oceans.

Sea Turtles

Sea turtles arose about 100 million years ago from terrestrial or fresh-water turtles. Currently only 7 species are extant although certain authors list Chelonia mydas agassizi as an eighth valid species. Sea turtles are found primarily along tropical coasts. However, some are also well-known for their long journeys across the oceans. Most species nest along the coasts of Central and South America or in the Caribbean, although some species occasionally travel as far north as Scandinavia.

Like many other reptiles, most if not all sea turtles seem to use temperature-dependent sex determination (TSD). TSD has been demonstrated for loggerhead (Caretta caretta), green (Chelonia mydas), leatherback (Dermochelys coriacea) and olive ridley (Lepidochelys olivacea) turtles. Since TSD is often sensitive to small changes in temperature, global warming may eventually affect sex ratios in these species and, together with their general vulnerabilities, have a dramatic effect on their reproductive rate and thus long-term survival.

Migration

Usually, sea turtles travel the seas on their own. However, during the nesting season they head towards their home beaches and may form large groups of turtles traveling together, even if they maintain distances of up to several hundred meters between individuals. The ability of sea turtles to find their original nesting sites has spurred considerable interest. Only recently it has been shown that sea turtles use geomagnetic sensing to orient themselves. While sea turtles appear to return to a certain geographic region, they not necessarily return to a specific. However, Kemp's Ridley does nest on only one location in Mexico, so it is especially endangered. It remains unclear though how precise other turtles are in returning.

Conservation

Most sea turtles are endangered or even critically endangered. The Olive Ridley Turtle (Lepidochelys olivacea) is vulnerable and for the Flatback Turtle (Natator depressus) there is simply not enough data, although in previous years it had been classified as "vulnerable" too. The IUCN Red List says about the Green Turtle (Chelonia mydas):

> "Analysis of historic and recent published accounts indicate extensive subpopulation declines in all major ocean basins over the last three generations as a result of overexploitation of eggs and adult females at nesting beaches, juveniles and adults in foraging areas, and, to a lesser extent, incidental mortality relating to marine fisheries and degradation of marine and nesting habitats. Analyses of subpopulation changes at 32 Index Sites distributed globally show a 48% to 67% decline in the number of mature females nesting annually over the last 3 generations."

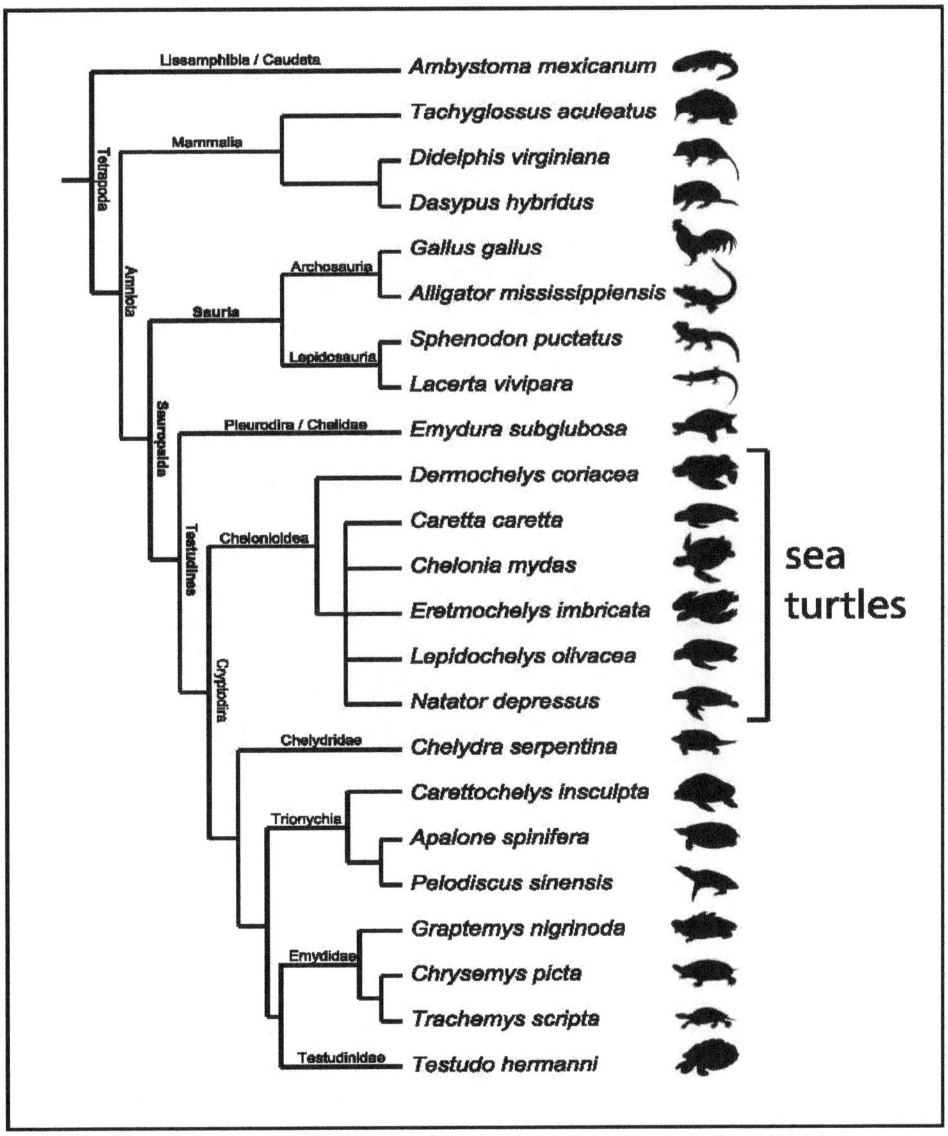

Figure: Phylogeny of sea turtles: (A) Phylogenetic relationships of amniotes with the position of sea turtles relative to other vertebrates.

Similar statements are true for all sea turtles: overexploitation and the destruction of nesting sites cause concern for these species, as tropical beaches are also in high demand for touristic reasons and easily exploited by locals.

Table: All species of turtles.

Species	Common name	IUCN Red List status
Caretta caretta	Loggerhead	Endangered
Chelonia mydas	Green Turtle	Endangered
Dermochelys coriacea	Leatherback	Critically Endangered
Eretmochelys imbricata	Hawksbill	Critically Endangered
Lepidochelys kempii	Atlantic Ridley	Critically Endangered
Lepidochelys olivacea	Pacific Ridley	Vulnerable
Natator depressus	Flatback Turtle	? (data deficient)

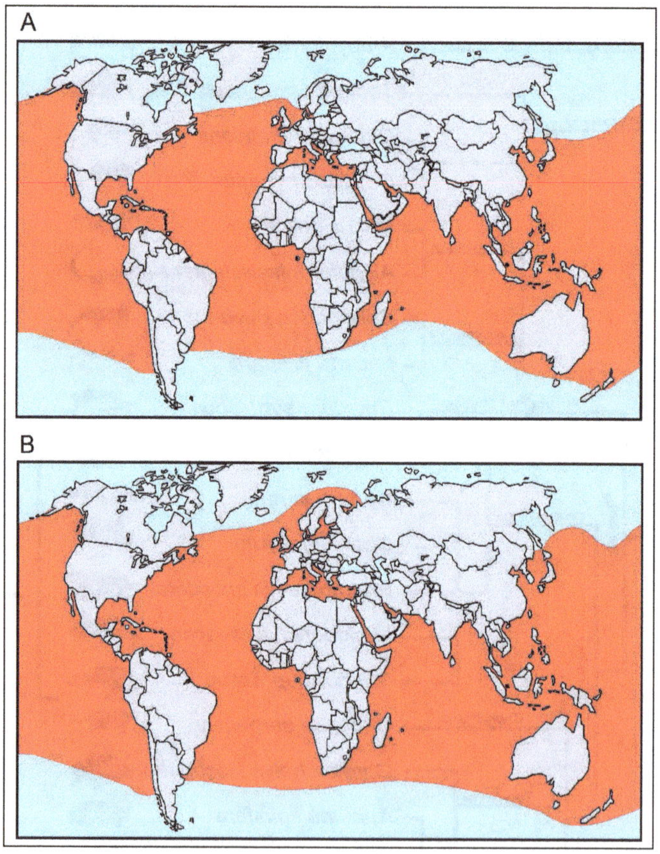

Figure: (A) Cheloniidae. (B) Dermochelyidae.

Other Turtles Living in Brackish Environments

Of the more than 300 species of turtles only seven are truly marine while about 50 species are fully terrestrial, belonging to the family of tortoises, the Testudinidae. The majority of the remaining species, and most of the world's turtles, are aquatic or semi-aquatic freshwater species. However, a few are associated with estuarine and other brackish water habitats in an environment that is neither marine nor fresh water.

Table: Turtles in brackish waters.

Genus	number of all species	brackish species
Malaclemys	1	1
Batagur	6	3
Carettochelys	1	1
Pelochelys	3	1

Included among the species that spend a portion or all of the year in estuarine habitats are the mangrove terrapins (Batagur affinis and B. baska) of south-east Asia and India, as well as the pig-nosed turtles (Carettochelys sp.) in southern New Guinea. Painted terrapins (Batagur borneoensis) of south-east Asia characteristically spend an even greater portion of their life cycle in estuarine and brackish waters, even laying their eggs on oceanfront beaches in the same areas as sea turtles. A few additional species of turtles, such as the giant softshell (Pelochelys cantori) of Asia, will enter estuarine brackish waters and even into full saltwater habitats temporarily, but the majority of their range includes freshwater habitats.

The only exclusively brackish water turtle in the world is the diamondback terrapin (Malaclemys terrapin), which is endemic to tidal creeks and salt marshes as well as the brackishportions of estuaries of the Atlantic and Gulf coasts in the United States, from Cape Cod to Texas. Diamondback terrapins in Florida are found in mangrove swamps. Two species of Batagur in Asia and the painted terrapin in Malaysia also occupy mangrove swamps. All of the species found in mangrove swamps also inhabit other brackish waters of coastal areas and most are associated to some degree with river estuaries, and the Asian species commonly enter freshwater systems as well. None of the turtle species in South America, Africa, or Europe has an affinity for brackish water conditions characteristic of the species noted above. However, many species of freshwater turtles that live in coastal areas around the world are known to enter brackish waters on occasion. The diamondback terrapin has a functional salt gland and can live indefinitely in fresh water or sea water. Other estuarine turtles have not been reported to have well developed salt glands comparable to those in the diamondback terrapin.

All turtle species that are brackish water inhabitants face severe conservation threats in all or part of their range because of the numerous environmental impacts on coastal systems throughout the world. Habitat degradation from pollution, sediment runoff, and other consequences of overdevelopment as well as mortality as targeted commercial species or as bycatch from the fisheries industry take an increasingly greater toll on these species.

Figure: Malaclemys terrapin: This species is the only terrestrial turtle with significant adaptions to coastal habitats.

Marine Crocodiles

None of the currently 23 species of crocodile is truly marine (Crocodylus raninus has been re-validated as 24th species only recently but this has not been universally accepted). However, at least one species, the Saltwater or "Estuary" Crocodile (Crocodylus porosus) is regularly found in brackish waters of south-east Asia and Australia. Other crocodiles have been found in tidal waters, such as Crocodylus johnstoni or C. acutus, but C. porosus is the only one that does show some adaptations to salt water. The saltwater crocodile is the largest living crocodile and thus the largest living reptile, reaching a total length of more than 6 meters. The species seems to be most closely related to C. siamensis which is not particularly associated with brackish waters.

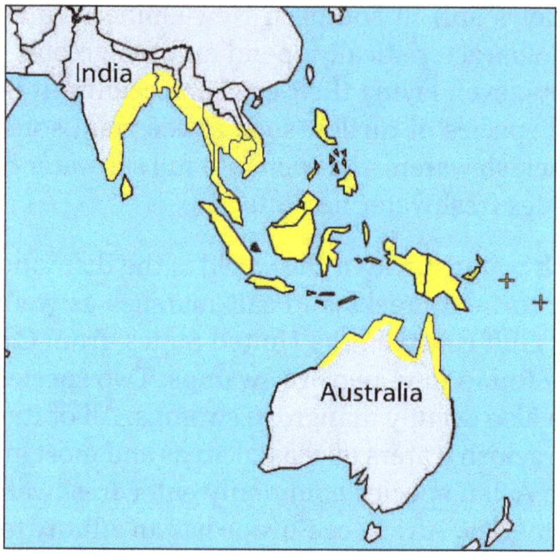

Figure: Distribution of the saltwater crocodile, Crocodylus porosus.
The range is shown in yellow. "+" symbols represent the Pacific islands
that are also inhabited by this species, including the Solomon islands and Vanuatu.

The saltwater crocodile has a high tolerance for salinity, being found in brackish water around coastal areas and in rivers. However, it is also present in freshwater rivers and swamps. Movement between different habitats occurs between the dry and wet season, and as a result of social status: juveniles are raised in freshwater areas, but eventually sub-adult crocodiles are usually forced out of these areas (used for breeding by dominant, territorial adults), into more marginal and saline areas. Subordinate animals unable to establish a territory in a tidal river system are either killed or forced out into the sea where they move around the coast in search of another river system. In recent years in northern Australia, saltwater crocodile populations in some areas have recovered to such an extent that increasing numbers are being forced further upstream into marginal habitat.

Crocodiles use their lingual salt glands to secrete excess salt ions. The morphology of these salt-secreting glands is highly conserved. These tissues are typified by their abundance of ion pumps, responsible for the maintenance of cellular electrochemical gradients through the movement of Na^+ and K^+ ions against their osmotic gradients. Crocodylus porosus possesses lingual salt glands which function to remove excess Na^+ and Cl^- ions accumulated as a consequence of living in a marine environment. However, other crocodiles, such as the Nile crocodile, seem to have similar glands, even though they may be not as active or efficient as those in the saltwater crocodile. Sea water are evidence for a marine evolutionary origin of crocodiles.

Sea Snakes

Besides the sea turtles, the sea snakes are the reptiles that are best adapted to marine environments. The most typical feature of a sea snake is the vertically flattened paddle-like tail, which is not found in any other terrestrial or aquatic snakes. Sea snakes occur in the tropical and subtropical waters of the Indian and Pacific Oceans from the east coast of Africa to the Gulf of Panama. However, two specimens of Pelamis platurus reported from Namibia indicate that the species may be extending its range into the Atlantic Ocean. Most species are concentrated in the Indo-Malayan Archipelago, South China Sea, Indonesia and the Australian region. Sea snakes inhabit shallow waters along coasts and around islands, river mouths and up rivers for more than 150 km and they have also been found in lakes in Thailand, Cambodia, the Philippines and Rennell Island in the Solomon archipelago. However, information on precise geographical distribution and abundance for each species is still lacking.

Figure: A typical sea snake: Hydrophis belcheri. Note the flattened tail as an adaptation to swimming in the open sea.

Notably, a few sea snakes are actually freshwater species, such as Hydrophis semperi, H. sibauensis, H. obscurus, and L. crockeri. However, all evidence indicates that these species have radiated into freshwaters independently from saltwater species.

Sea Snake Bite

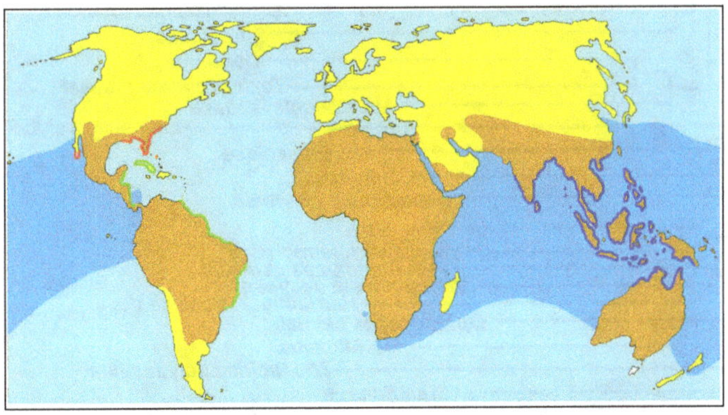

Figure shows Distribution of marine snakes, Terrestrial distribution represents terrestrial elapids (brown), marine distribution represents sea snakes, i.e. the subfamily Hydrophiinae of the Elapi-dae (blue). Dark blue: homalopsid snakes along the Asian and Australian coasts. Red: North-American Natricidae, green: neotropical Dipsadidae.

Sea snake bite is the cause of human fatalities. The typical victim is a fisherman handling gape nets or down nets, sorting fish (and sea snakes) on board a trawling boat or dragging a net by wading in muddy coastal waters or in river-mouths. Some sea snakes are gentle, inoffensive creatures that bite only under provocation, but other species are much more aggressive (e.g. Aipysurus laevis, Astrotia stokesii, Enhydrina schistosa, Hydrophis elegans, H. macdowelli, H. major, H. ornatus and H. ocellatus). As a general rule all sea snakes must be handled with great caution: however, it is worth mentioning that when sea snakes do bite, they do not always inject much or any of their venom, so that only trivial severity of poisoning will be recognizable.

Taxonomy

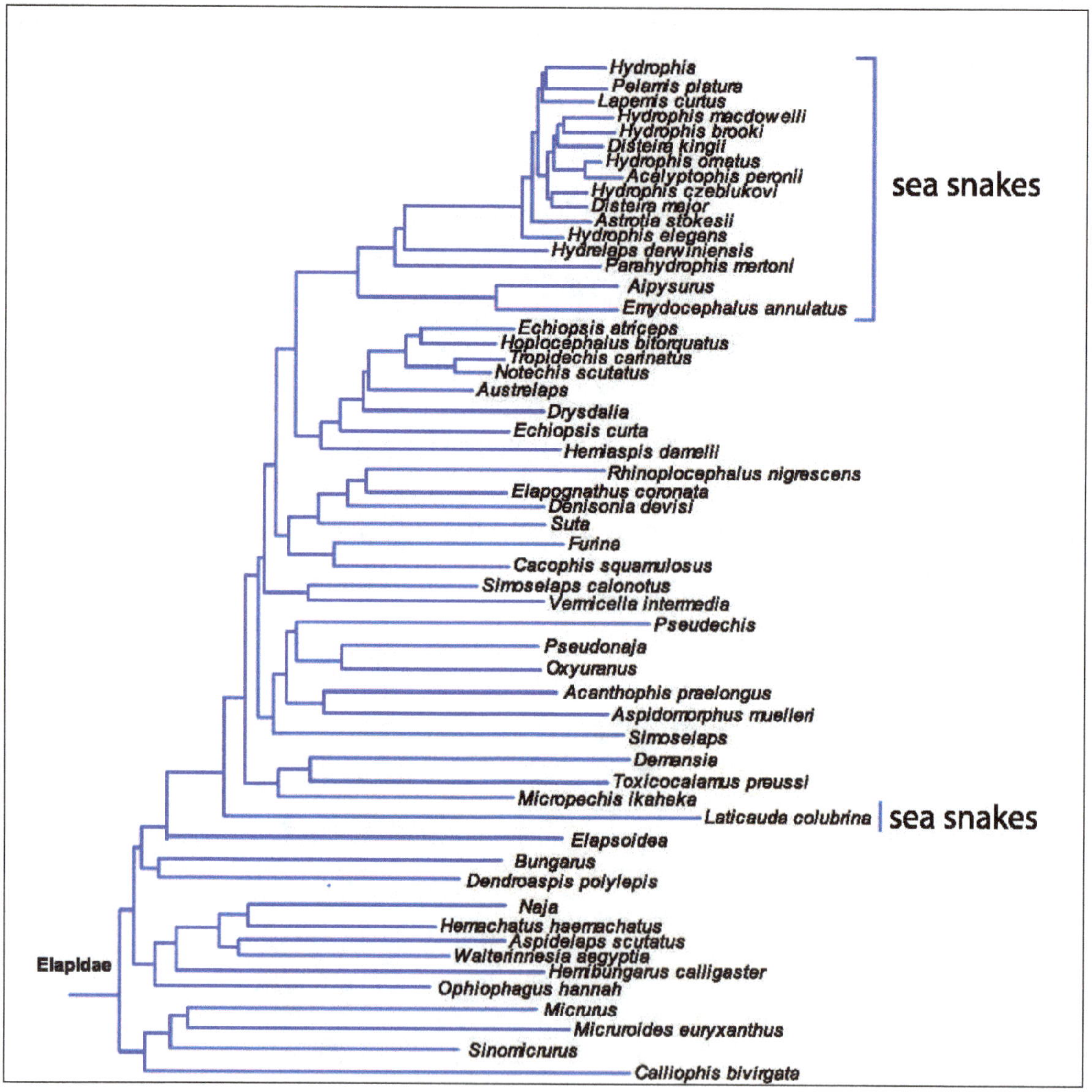

Figure shows Sea snakes and their relationship to other Elapid snake genera. Note that the the "true" sea snakes (Hydrophiinae and Laticaudinae) are not closely related but nest within the terrestrial elapids. Generic names represent multiple species. The relationships among the many species of Hydrophis have not been resolved.

The taxonomic status of the sea snakes is still under review and no general agreement exists at the moment. Traditionally sea snakes have been regarded as belonging to one family, Hydrophiidae, with Laticauda as the most primitive genus. However, more recent results support the position that Laticaudinae and Hydrophiinae (true sea snakes) have evolved from different terrestrial elapids. The combined morphological and molecular results by Scanlon and Lee support that sea kraits (Laticauda) and Solomon Islands elapids are basal to the remaining Australian terrestrial elapids and true sea snakes (Hydrophiinae). The Australian Elapids and true sea snakes include three main lineages: a large-bodied oviparous lineage, a small-bodied oviparous lineage, and a viviparous lineage which also includes the true sea snakes. The results by Castoe et al. indicate that Laticauda is closest to some Asiatic elapids. However, these authors suggest to interpret these results cautiously because of long branch attraction. Sanders et al. indicate that Laticauda is the sister group to all other hydrophiines (oxyuranines), which is also consistent with a classic morphological feature: all oxyuranines have reduced the choanal process of the palatine and lost the lateral process, permitting novel jaw movements.

Some results indicate that true sea snakes (Hydrophiinae) can be separated into two totally different groups, indicating that sea snakes perhaps have evolved three times from terrestrial elapids. However, this hypothesis is not supported by other recent results and the latest molecular evidence suggests that the true sea snakes (Hydrophiinae) are a very young monophyletic group, perhaps only 8–13 Myr old. On the genus level at least three completely different taxonomic hypotheses exist, resulting in much confusion in both scientific and non-scientific publications. The most thorough work on sea snakes is by Smith, 1926, whose taxonomy has been used by most authors during the past 85 years. In 1972 McDowell, based on a few specimens from selected species, suggested separating the genus Hydrophis into 3 different subgenera based on morphological data. McDowell also changed some genera using the same arguments, which were not based on a phylogenetic analysis, but more on morphological similarity. Kharin has suggested raising some of McDowells' species groups to genus level, however, Kharin also did not produce any phylogenetic analysis for these groups. Recently, Wells suggested changing the nomenclature in the Australian sea snake species also without any phylogenetic analysis, thus creating new problems for the layman to use valid generic names for the groups. A couple of papers using phylogenetic methods do not support the hypothesis by McDowell. Because of sea snakes' extremely venomous bite it is important to have correct taxonomic assignments and thus correct names because physicians often apply anti-venoms based on the species of snake. As long as no thorough phylogenetic analysis is available for the group, we suggest using Smith's classification, with a few corrections that have been generally accepted.

Identifying sea snakes to species level is very difficult; especially the genus Hydrophis shows great intraspecific variation making it difficult to use only external characters for their identification. The 70 species recognized here follow Golay et al. and David & Ineich with a few new species added or resurrected. A complete species list can be found in when searching for the two subfamilies, Hydrophiinae and Laticaudinae, respectively.

At a higher taxonomic level sea snakes are the closest relatives of terrestrial elapids, which include some of the most venomous snakes in the world (e.g. brown snakes, taipan, death adder, cobra, krait, mambas). Sea snakes or aquatic elapids and terrestrial elapids are collectively known as proteroglyphous snakes because of the position of the poison-fangs in the frontal part of the upper jaw (maxillary bone).

Feeding and Breeding Biology

Most sea snake species feed on fish that are close to the bottom or sedentary; a few prefer fish eggs (Aipysurus eydouxii, and the genus Emydocephalus; some specimens of A. fuscus also contained fish eggs) and at least Aipysurus laevis, Enhydrina schistosa and Lapemis curtus have been found with crustaceans and mollusks in their stomachs. The Laticauda group feed mostly on eels burrowed in sand at the bottom of the sea and in reef crevices. Using sea snakes to capture undescribed eel species has been shown to be very effective because the eels are extremely secretive in their habits giving no chance to collect them using traditional methods.

All sea snakes produce living young (viviparous) except for the genus Laticauda which is egg laying (oviparous). The annual reproductive cycles are synchronous between males and females and they reproduce every second year with a clutch size that increases with the size of the female.

By-Catch and Commercial Use of Sea Snakes

Sea snakes are not only of interest because of their poison, but also in connection with the commercial exploitation of reptile skin, organs and meat. Some species are accessible in great numbers (e.g. Laticauda spp., Lapemis spp. and some Hydrophis spp.) and are not protected by CITES (Washington convention).

Since at least 1934 sea snake meat and skins have been used commercially in the Philippines, but also in Australia. In Japan, Taiwan, Thailand and Vietnam sea snakes have been collected commercially. Local protection of sea snakes has been necessary to stop over-exploitation in the Philippines. In Australia commercial sea snake fisheries and by-catch have been investigated during the past 15 years. However, most sea snake fisheries in the Indian Ocean and in the Pacific are not reported in the literature and are beyond control of the local governments.

Species Distribution and Density

Many of the more than 60 species of sea snakes have a broad distribution in both the Indian Ocean and the Pacific Ocean. Species such as Acalyptophis peronii, Aipysurus eydouxii, Astrotia stokesii, Enhydrina schistosa, Lapemis curtus, Laticauda colubrina, L. laticaudata and many Hydrophis species have been collected in both the Asian and the Australian region and are abundant in these areas. Other species are much less well known and some species are only collected from very restricted areas and therefore much more vulnerable for change in the environments. One of the rarest species is Hydrophis parviceps which is known from only two specimens collected in a very limited area in the southern Vietnamese part of the South China Sea. Hydrophis sibauensis is a recently described species which has been collected more than 150 km into rivers of Borneo and is only known from 3 specimens. This is also the case for another recently described species, Hydrophis laboutei which is only known from two specimens collected at Chesterfield Reefs (New Caledonia). Other species with a limited distribution are therefore highly vulnerable to all kinds of environmental changes, including Aipysurus fuscus, A. apraefrontalis and A. foliosquama which have only been collected in the north west part of Australian coral reefs. Recent surveys indicate that there has been a drastic decline of specimens in this areas over the last 10–15 years and pers.comm. Michael Guinea). Hydrophis semperi and Laticauda crockeri, both species only known from Lake Taal in the Philippines and Lake Te-Nggano in Rennell Island, respectively, also represent species that have a very restricted distribution and therefore are highly vulnerable to changes in the environment.

Our own (ARR) investigations in the Gulf of Thailand and part of Borneo indicated that the sea snake fauna in this area has also declined. In the Gulf of Thailand, especially populations of species going up rivers have disappeared (e.g. Hydrophis torquatus and H. klossi), but also in Borneo the number of specimens collected previously in great numbers inside river mouths appears to have declined. Trawl fisheries in the Andaman Sea (e.g. around Phuket, Thailand) also got fewer specimens and the fishermen also mention that the size of sea snakes they got in former times was much larger than the ones they get now. A small survey in Cambodia (ARR) in the year 2000/2001 indicated similar results to those found for Borneo and Thailand.

Conservation of Sea Snakes

The main threat to sea snakes is the cultural indifference to conservation issues by locals and, as a consequence, their commercial exploitation. Only raising awareness may reduce this kind of threat. Another problem in trying to help the sea snakes to survive is the very limited knowledge on their biology, especially in Asia. It is very important to get much more information on the biology in formulating management plans. For the moment we still miss information on breeding cycles, by-catch and mortality, growth rates, population density, sexual maturity and taxonomy in most areas. The effects of the exploitation or/and by-catch on the sea snakes are almost unknown except from the Philippines and Australia. Some populations may already be in danger of extinction. The only way to have a sustainable yield is by monitoring and controlling by-catch and commercial catch of sea snakes, giving local governments a chance to intervene before a catastrophic collapse of local populations occur. However, to limit exploitation of the most common species sustainable and to protect the endangered ones, we need to have much more biological focus on the group.

One cause for the disappearance of sea snake species from rivers is at least in part due to the great problems with the ongoing pollution in many rivers in both Asia and Australia. Also the information on breeding areas for most sea snake species is unknown, but there is no doubt that also habitat destructions (e.g. mangrove clearings and water crafts) have a negative influence on sea snake populations.

A first step in sea snake conservation is to distinguish the many species from each other, which is not an easy task, and with the global decline of taxonomists it may be more important to focus initially on the entire group. The next step is to get more knowledge of sea snake biology and then to focus on the bycatch and management plan to protect the endangered species and harvest only the more common ones. Concerning the more global problems of pollution and habitat degradation, we have to put more pressure on politicians and hope that they will come up with solutions for the benefit of both humans and sea snakes.

Other Snakes Found in Marine and Brackish Environments

Evolution is continually tinkering with snake populations that live in coastal areas adjacent to brackish and salt water environments. Besides the true sea snakes, many terrestrial and arboreal species have learned to exploit marine resources by foraging in the intertidal zone at low tide or from the branches of mangroves, while some freshwater species have adapted to life in brackish water, sometimes enter the ocean, or live there permanently.

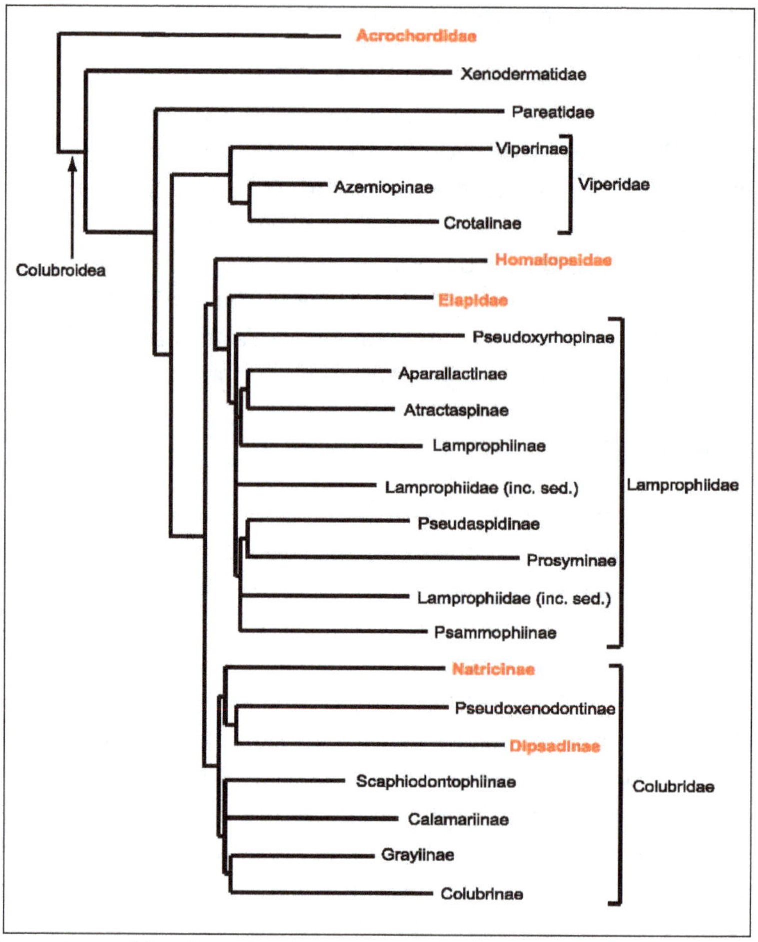

Figure: Several families of snakes have independently adapted to saltwater. Families with species that live in brackish or marine environments are shown in red.

Table: Non-Elapid snakes that use brackish water.

Genus	Total species	Species in brackish or marine water
Acrochordus	3	All (3)
Agkistrodon	4	1
Bitia	1	All (1)
Cantoria	1	All (1)
Cerberus	3	2+
Crotaphopeltis	6	1
Djokoiskandarus	1	All (1)
Enhydris	24	1
Farancia	2	All (2)
Fordonia	1	All (1)
Gerarda	1	All (1)
Grayia	4	1
Helicops	15	2+

Hydrops	3	1
Liophis	39	2+
Myron	3	All (3)
Natrix	4	3+
Nerodia	10	4+
Regina	4	1
Stegonotus	9	2+
Storeria	4	1
Thamnophis	31	2
Tretanorhinus	4	2

The list does not include arboreal and terrestrial species that use mangrove forests or terrestrial species that occasionally enter water, or forage in the intertidal zone. A "+" indicates there are probably more species in this genus that use brackish water.

All three living species of the ancient Acrochordidae use a combination of aquatic environments ranging from freshwater to sea water, and have probably been living in coastal ocean habitats for most of their 90.7 million year history. Acrochordids, particularly Acrochordus granulatus have some of the most specialized morphology and physiology for life in saltwater, including the greatest capacity to store oxygen found in any vertebrate. These are low metabolism snakes, feeding and reproducing infrequently and incapable of active swimming for more than a few minutes. Acrochordus arafurae females ambush prey in deep water while males actively search for prey in shallow water. They have a constriction-like behavior for holding prey between body loops during ingestion, and enlarged keels on their scales aid in holding slippery fish. Additionally each scale contains a mechanoreceptor that may be used to locate fish in turbid water.

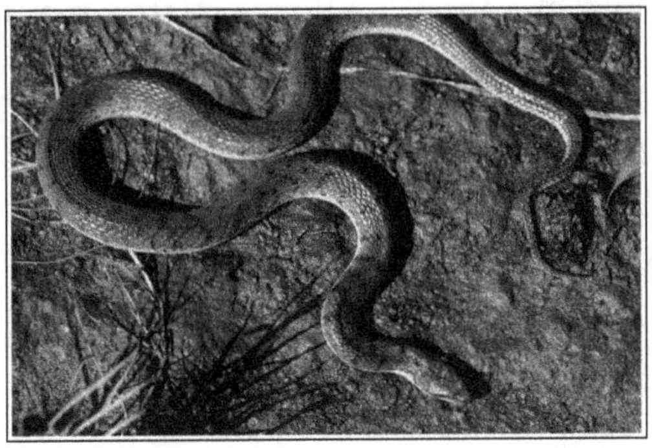

Figure: A homalopsid snake (Cerberus rynchops).

The Asian and Australasian Homalopsidae, ironically sometimes called the "freshwater snakes," have species that live in coastal habitats such as mangrove forests and salt marshes. Homalopsids inhabit life zones that range from the fossorial Brachyorrhos albus and the semi-terrestrial Enhydris plumbea, to Erpeton tentaculatus that is all but incapable of leaving their freshwater habitats, to species inhabiting coastal marine environments (Bitia hydroides, Cantoria violacea, Myron sp., Fordonia leucobalia, Enhydris bennettii). The most widespread and successful brackish water homalopsids are the bockadams (Cerberus sp) which are distributed from the vicinity of Mumbai, India in the east to Palau, Micronesia in the west, and range southward into the Indonesian Archipelago, New Guinea, and northern Australia. Like some of the Hydrophis and

Laticauda, the Lake Buhi Bockadam (Cerberus microlepis) is a freshwater species, derived from its nearby salt water dwelling relative, Cerberus rynchops in the Philippines. While most of the aquatic homalopids are piscivorous, three species, all members of the same clade specialize in feeding on crustaceans (Cantoria violacea, Fordonia leucobalia, and Gerarda prevostiana) in near shore habitats.

An assemblage of mangrove homalopsids studied in Singapore was dominated by the piscivorous Asian Bockadam, Cerberus rynchops (73% of total snakes), while the other species were crustacean specialists and less common. Gerard's Mud Snake, Gerarda prevostiana (16% of the total snakes) feeds exclusively on recently-molted crabs; while the Crab-eating Snake, Fordonia leucobalia, (10% of total snakes) specializes in feeding on hard-shell crustaceans. The most uncommon Singapore homalopsid was Cantor's Mud Snake, Cantoria violacea, (2% of total snakes) and it too has a specialized diet, feeding on Alpheus snapping shrimp (Decapoda, Alpheidae). The crustacean-eaters were often observed in association with mud lobster mounds constructed by Thalassina anomala (Decapoda: Thalassinidae). The snakes were nocturnal and active throughout the night. Gerard's Mud Snake increased activity during spring tides, but the other species did not. Three male Crab-eating snakes were tracked using radiotelemetry: they rarely moved and when they did it was for short distances (1.8 to 14.0 m). As might be expected for tropical aquatic snakes, the body temperatures were very stable (26.3–29.0uC) and consistently higher than the microhabitat temperatures. Two of the radio tracked crab-eating Snakes made extensive use of mud lobster mounds.

Snakes of the family Natricidae (or subfamily Natricinae) make up about 30 genera and 210 species. They occur in both hemispheres and in both temperate and tropical environments, but they appear to have originated in Asia, and dispersed into Europe, and North America. Many of these snakes are semiaquatic and some inhabit brackish water. In North Africa and Europe at least three species use coastal habitats: the Viperine Snake, Natrix maura, the Grass Snake, N. natrix sicula, and the Dice Snake, N. tessellata. North American brackish water natricids use mangrove and salt marsh habitats include the Mangrove Water Snake, Nerodia clarkii, in the southeastern USA, and the Baja Garter Snakes, Thamnophis valida. Both these species are piscivorous and like some freshwater natricids, Nerodia clarkia juveniles lure prey with their tongue. The Salt Marsh Brown Snake, Storeria dekayi limnectes is less aquatic than the previous species, and it feeds on soft bodied invertebrates.

In the Neotropics, the family Dipsadidae (or subfamily Dipsadinae) has more than 92 genera and >700 species, a few of which use brackish and salt water habitats (Helicops, Hydrops, Liophis, and Tretanorhinus) to varying degrees. For the most part these snakes are poorly known and their brackish water habits are known only from anecdotal observations. Similarly, at least two aquatic African snakes, of uncertain lineages (Grayia smythii and Crotaphopeltis hotamboeia) occur in the brackish water of mangrove forests as well as freshwater. There is another group of snakes that use marine resources that is worthy of note. These species forage in mangroves, salt marshes, and intertidal zone without actually spending much time in the water. The most specialized is perhaps the Burmese Vine Snake, Ahaetulla fronticincta (family Colubridae) which hunts gobies from branches over hanging the water. The Southeast Asian Mangrove Pitviper, Cryptelytrops purpureomaculatus (Viperidae) also hunts from the mangrove branches but its diet is unknown. There are others, for the most part habitat generalists with populations in or adjacent to marine habitats and include Python molurus, Python reticulatus (Pythonidae), Ophiophagus hannah (Elapidae), and Boiga dendrophilia (Colubridae) – all medium to large snakes that forage for food in coastal habitats. Other snakes forage in the intertidal zone: Coluber anthonyi (Colubridae); Crotalus mitchelli and Crotalus muertensis (family Viperidae, subfamily Crotalinae).

Thus, of 34 lineages of snakes (families and subfamilies), four (Acrochordidae, Homalopsidae, Dipsadidae, and Elapidae) contain most of the species adapted for marine environments, while other clades have relatively few, or no, species adapted for the saline water. As more herpetologists investigate mangroves and salt marshes a more complete picture of brackish water snakes will emerge.

Marine Iguanas

Lizards are the most speciose and diverse group of reptiles, with almost 5,500 species (60% of all reptiles). Nevertheless, only a few species have ventured into the oceans. The marine iguanas of the Galapagos Islands are the most aquatic of the lizards, but bask and reproduce on land and are subject to terrestrial predators.

The Galapagos archipelago arose through volcanic activity 960 km off the coast of Ecuador. Lizards and other animals probably reached the island on rafts of vegetation washed down the rivers of western South America. The rafts may have included juveniles, adults, or eggs. The Humboldt and El Niño currents would allow for such rafting to originate from the western coast of South or Central America. Other species, such as rats, have been introduced by humans only about 100 years ago.

Besides seven species of smaller iguanids of the genus Microlophus four large species of iguanas inhabit the Galapagos Islands: three species of land iguanas (genus Conolophus) and the related marine iguanas (Amblyrhynchus cristatus). Because the land iguanas are endemic to only a few islands, they are threatened by predation of introduced mammals. However, recent measures to control and protect their habitats seem to have stabilized their populations. In addition to these anthropogenic factors, the archipelago is subject to strong seasonal and annual variations in environmental conditions, so that a combination of factors could turn out to be detrimental.

The marine iguana, Amblyrhynchus cristatus, and the land iguanas of the genus Conolophus are all about 1.2 m in length. Amblyrhynchus inhabits virtually all islands of the archipelago, given its ability to disperse to various islands. As a result, they are less prone to extinction.

Although land and marine iguanas are very different in appearance, they are closely related. In fact, several instances of hybridization between the two genera have been reported. Hence it is likely that they both evolved relatively recently from a land iguana that came from the South American mainland.

Marine iguanas feed exclusively on marine plants. While they spent a considerable time in the water foraging, they have not completely adapted to marine life. For instance, they still have to nest on land and also bask on land to reach their optimal body temperature which rapidly declines in the rather cool ocean water. Nevertheless, considerable selection pressure has resulted in several adaptations to their marine lifestyle such as a flattened tail and limited webbing of all four feet, supporting swimming. Powerful claws help them to hold on to rocks in the heavy sea. Marine iguanas also have reduced the number of heart beats per minute from about 43 on land to 7 to 9 while diving, as do several other reptiles. Finally, both Conolophus and Amblyrhynchus possess nasal salt glands, similar to those found in other reptiles that have high dietary salt intakes. Interestingly, neither species has the capacity to produce hyperosmotic urine. Thus, the marine iguana has the highest known extracloacal excretion rate of Na, Cl, and K of any reptile.

Marine Mammals

Marine mammals occupy a wide range of marine and some freshwater habitats around the world. They have been used by humans for millennia for food and to obtain other products. Marine mammals consist of cetaceans (whales, dolphins and porpoises), pinnipeds (seals, sea lions and walruses), sirenians (dugongs and manatees), mustelids (sea otters and marine otters) and the polar bear – 130 or more species in total, including several that range into fresh water and some that exclusively occupy rivers and lakes. Human interactions, both direct and indirect, have negatively affected most marine mammal species at least to some degree. Historically, industrial harvesting greatly reduced the abundance of many populations. Although the intensity of such exploitation has declined in recent decades for many species, humans continue to use certain species of marine mammals in some places for food, skins, fur, ivory and increasingly as bait for fisheries. In addition, human activities continue to reduce the availability and quality of marine mammal habitats and cause substantial numbers of marine mammals to die incidentally as a result of entanglement or entrapment in fishing gear and from being struck by vessels. Mitigation of ongoing and future threats to marine mammals from human activities requires improved knowledge of those human activities and of the animals' ecology, behaviour and habitat use.

Changes in Biological Diversity

Globally, certain species of marine mammals have become extinct over the last several centuries, i.e., Steller's sea cow, the Japanese sea lion, the Caribbean monk seal. The Yangtze River dolphin (baiji) is also likely to have become extinct although there has been one unconfirmed sighting in 2005. In addition, many populations have been reduced to remnant status, such that they no longer play a significant role in the ecosystem, i.e., they are functionally extinct. For example, right whales have essentially disappeared from the eastern North Atlantic, and walruses are gone from their former strongholds in south-eastern Canada (Gulf of St. Lawrence and Sable Island). The loss of marine mammal diversity due to actual and functional extinctions has had significant effects on marine ecosystems at varying scales from local to ocean-basin-wide.

Magnitude of Changes

Besides the above-mentioned extinctions and extirpations, the numerical abundance (and biomass) of many other marine mammal populations has been greatly reduced. All of the commercially valuable great whales were depleted by whaling. A good example is the world's largest animal, the blue whale. Within the 20th century more than 360,000 Antarctic blue whales were taken, leaving a remnant population of only hundreds of animals and a current abundance of roughly 5 per cent of original numbers. Some populations of virtually all types of marine mammals have been depleted to low levels; for example, southern and northern hemisphere populations of fur seals and elephant seals were all depleted by harvesting for either their fur or their oil (Busch, 1985). Similarly, many populations of sirenians and sea otters have been reduced to small fractions of historic levels.

Although the abundance of almost all populations of marine mammals has been reduced by human activities, a number of them have recovered since they were protected from deliberate exploitation. Eastern Pacific gray whales, northern elephant seals, eastern Steller sea lions, humpback whales and some populations of right whales provide some of the most clear-cut examples. Certain populations of those species, however, have not recovered and remain at small fractions of historic levels (e.g., western Pacific gray whales, Arabian Sea humpback whales, western Steller sea lions, North Atlantic right whales, and southern right whales in some areas).

Population Trends or Conservation Status Aggregated at Global Scale

Because of the great diversity of marine mammal species and their habitats, it is difficult to characterize their conservation status and population trends in the aggregate or at a global scale. Many of the marine species suffered major declines over the past several hundred years as a result of commercial hunting. Massive changes in human demography and economy have unquestionably affected the environmental carrying capacity, particularly in coastal regions, and as a result less suitable habitat (including forage base) is available to support marine mammal populations. These changes make 'full' recovery unfeasible for some species.

Major Taxonomic and Geographic Subdivisions

Large Whales

Commercial exploitation began as long ago as the 10th century, and by the late 19th century intensive whaling had caused severe depletion, and even near-extinction, of some species and populations. Industrial mechanized whaling in the 20th century led to further major declines. Some populations of large whales have been recovering in recent decades: for example, humpback whales globally, blue whales in some regions and southern hemisphere right whales when treated as a single group. At the same time, many populations have failed to recover to anywhere near their original abundance. For example, right whales are effectively extirpated from the eastern North Atlantic, and are only barely surviving in the eastern North Pacific the eastern South Pacific, and around New Zealand.

Pelagic Dolphins and Porpoises

Pelagic (off-shore) dolphins are generally less susceptible to human interactions than many marine mammals because they are relatively small, of little commercial importance, wide-ranging (some species are also extremely numerous) and live far from most human activities. There are potential interactions with military activity, and clear interactions with fishing gear in the eastern tropical Pacific. Several dolphin species in that region have been significantly affected by humans because they have symbiotic relationships with other pelagic animals that are of commercial interest, notably yellowfin tuna. Purse seine fishermen learned to chase and encircle these dolphins to facilitate the capture of tuna, and in the past large numbers of dolphins drowned in the nets. This mortality greatly reduced the abundance of several dolphin species in the latter half of the 20th century. Due to international efforts, fishing methods have changed and the by-catch has been reduced significantly.

Coastal and Estuarine Dolphins and Porpoises

Coastal dolphins and porpoises are hunted for food and bait by a number of States. Furthermore, they often die accidentally in fishing gear in coastal commercial and artisanal fisheries. This has contributed to declines in abundance, at times to critical levels. For example, vaquitas in the Gulf of California (Mexico) declined by 57 per cent from 1997-2008, and continue to decline. In New Zealand, the Māui dolphin (a subspecies of Hector's dolphin) has declined to very low levels. In both cases, there have been management interventions aimed at minimizing direct mortalities and entanglements. However, under current levels of direct mortalities and entanglements, these trends are expected to continue, with both the vaquita and the Māui dolphin at risk of becoming extinct within the next few decades.

Several species of small cetaceans live in estuaries and in large rivers, often in close proximity to humans, and are threatened by artisanal fisheries, pollution, and coastal development. Two species of South American river dolphins are relatively widespread and at least locally common in Amazonia and Orinoquia: the tucuxi and the boto. The endemic blind river dolphins of the South Asian subcontinent still inhabit large portions of the Indus, Ganges, Brahmaputra and Karnaphuli rivers. Although they are locally common in a few places, their numbers and ranges have declined markedly over the past century. Three relict populations of 100 or fewer Irrawaddy dolphins persist in the Ayeyarwady (Irrawaddy) River of Myanmar, the Mahakam River of eastern Borneo, Indonesia, and the Mekong River of Cambodia and Laos. All three populations are at high risk of extirpation from a variety of human activities. The baiji (Lipotes vexillifer), endemic to the Yangtze River, (China), is likely to have become extinct in the early 21st century, and the Yangtze population of narrow-ridged finless porpoise (Neophocaena phocaenoides asiaeorientalis) is declining rapidly, probably due to mortality in fishing gear, vessel strikes and habitat degradation.

Pinnipeds

Populations of many species of commercially exploited pinnipeds were greatly reduced in abundance from the 18th to the 20th century. They were targeted mainly for the oil from their insulating blubber and for their fur. Some species, such as the Caribbean monk seal and the Japanese sea lion, were driven to extinction, and others, such as the Mediterranean monk seal, as well as fur seals in many areas, were reduced to very low numbers. The hooded seal in the Greenland Sea has been hunted down to 10-15 per cent of its original population. However, some regional and ocean-basin-wide pinniped populations have been recovering, such as the harp seal in Canada and grey and common seals in the United Kingdom of Great Britain and Northern Ireland. In some cases they have been recolonizing habitat where they had long been absent, for example, the New Zealand fur seal has expanded into former habitats in New Zealand and Australia. In other cases populations have increased to historical levels, such as the Antarctic krill-eating seals.

Sirenians

Manatees and dugongs are the only totally aquatic herbivorous mammals, feeding primarily on submerged vegetation. Sirenians have a global range in about 90 tropical and subtropical countries. They have been and continue to be hunted, trapped, and netted for food, and human activities of many kinds have modified or otherwise degraded their habitat. Overall, populations have been greatly reduced in terms of both abundance and range. One species discovered in 1741, Steller's sea cow, was hunted to extinction in less than a quarter of a century after its discovery. The present status varies regionally.

Mustelids

Two species of mustelids are truly marine: the sea otter (Enhydra lutris), along the Pacific coast of North America and East Asia, and the marine otter (Lontra felina), along the Pacific coast of South America. Both species are very coastal and they have been extensively exploited for their pelts. The sea otter has recovered substantially in many areas of its original distribution, whereas the marine otter has not, as it continued to be harvested through much of the 20th century. Its contracted range is fragmented due to various kinds of human interaction.

Polar Bears

Polar bears are endemic to high latitudes of the northern hemisphere. They have a circumpolar distribution and depend on both sea ice (for hunting pinnipeds and for denning and reproduction in some areas) and land (for hunting hauled-out pinnipeds such as walruses, as well as for denning and reproduction). Most populations have been subjected to extensive killing, at least historically, mainly for meat, hides, and sport, and to protect human life and property. Commercial hunting is now prohibited, although substantial numbers of bears are killed every year (legally) by aboriginal people in Alaska (United States of America), Canada and Greenland (Denmark). The main long-term and range-wide threat to polar bears is the projected loss of sea ice habitat associated with climate change. Sea ice provides essential breeding habitat for ringed seals, the principal prey of polar bears, and the bears rely on ice as a hunting platform to gain access to the seals. Limited access to food leads to mobilization of fat reserves and the release of hazardous substances stored in fat, which are transported by the blood circulation to vital organs such as liver and brain. High levels of contaminants have been associated with negative health effects (e.g. enzyme activation, hormonal disturbance and weakened immune systems) in polar bears in Svalbard (Norway), Greenland (Denmark), and Hudson Bay, Canada.

Special Conservation Status Issues

Sources of Conservation Information

International Whaling Commission: The conservation issues associated with large whales and, to a lesser extent, small cetaceans are addressed in the reports of the Scientific Committee of the International Whaling Commission. Besides directed hunting, these reports increasingly have addressed other topics of concern, including habitat degradation, ship strikes, climate change, fisheries by-catch, noise disturbance, and ecological interactions.

International Union for Conservation of Nature (IUCN): The IUCN Red List is an authoritative and regularly updated source of conservation information for all species and subspecies of marine mammals and for many threatened subpopulations. This information is easily accessible. The authority for listing determinations resides within the relevant Specialist Groups (SG) of the IUCN Species Survival Commission: the Pinniped SG, Cetacean SG, Sirenian SG, Otter SG and Polar Bear SG.

Potential Value of Management Methods

Regulation of directed and indirect takes: The most obvious approach to management of human activities that result in the killing of marine mammals is to control directed (i.e. deliberate) takes. Numerous examples demonstrate the effectiveness of this approach, such as the strong recoveries of some severely depleted populations following protection from whaling, sealing, or other forms of deliberate exploitation. The status of marine mammal populations, especially in relation to the effects of directed takes, is usually measured in terms of the size of populations relative to historical levels. The likelihood of population recovery after takes are controlled depends on the survival of a viable nucleus of individuals in the population at the time of protection. Furthermore, suitable habitat, including food resources, needs to exist to support recovery.

In some cases, especially when a population has been reduced to low levels, it is also necessary to manage sources of incidental (i.e. unintentional) mortality, injury, and disturbance due to other human activities. This can include requiring the use of less wasteful fishing practices and less by-catch-prone fishing gear and the regulation of ship traffic to reduce the incidence of vessel strikes. Techniques to reduce by-catch include the use of acoustic deterrents, gear modifications, time-area closures, and gear switching, for example, from gillnet to hook-and-line or traps/pots. Some of these approaches have been the subject of considerable research. Other anthropogenic threats may also need to be addressed, including reduction of noise from anthropogenic sources, and reduction of contamination by toxic substances (including oil) and of marine debris (including discarded or derelict fishing gear, and plastics of all sorts). The effects of tourism, for example whale and pinniped watching, in modifying behaviour through close approach, particularly to breeding and nursery areas, may also need to be addressed. In any event, successful marine mammal management requires selecting and implementing suitable methods of enforcement.

Marine Protected Areas: Protecting marine mammals in specific areas (e.g., feeding, breeding, and resting areas) can sometimes be effective for addressing certain threats. Understanding the life history of the species, the degree of localization of the threats and the needs and interests of local human populations is critical to the design of area management. Effective protection depends on devising plans that consider local community needs, as well as establishing effective control and monitoring. Furthermore, marine mammals tend to be highly mobile, and some species migrate across multiple ecosystems and even entire ocean basins, so protection may be needed in more than one area. There is considerable interest in spatial management to protect and conserve marine mammals through the establishment of parks, reserves and sanctuaries.

Key Pressures Linked to Trends

Direct Removals

There are various types of directed takes, including commercial harvests, scientific sampling, captive display, and subsistence. Some of these are under management while others are unauthorized and unmanaged. The nature of direct removals varies among major marine mammal groups.

There is a very long-standing, but relatively small (811 whales over four centuries), capture of northern bottle-nosed whales in the Faeroe Islands (Denmark), recorded from at least the sixteenth century. Coastal dolphins continue to be taken not only in the Solomon Islands, but also in a large scale drive fishery in Japan. Dolphins are hunted with harpoons in various countries for human consumption or for bait, especially in shark fisheries. The consumption of by-catch as well as deliberately taken cetaceans and sirenians is a growing concern in Africa, Asia, and some parts of Latin America. In the Arctic, narwhals and belugas continue to be hunted by aboriginal people in Canada, Greenland (Denmark), the Russian Federation and the United States. Marine mammals are also harassed and sometimes deliberately killed around aquaculture facilities in many parts of the world.

Whales are taken by several States for commercial, aboriginal and scientific purposes. The nature and regulation of these takes are described by the International Whaling Commission. Cetaceans occasionally strand on the shore in large groups: so-called mass strandings. The cause of this behaviour is poorly understood, but it has occurred for millions of years. Unless a population has been depleted by other causes, mass strandings are unlikely to be a major threat in themselves.

Pinnipeds – including walruses and numerous seal species – are hunted in large parts of the Arctic and sub-Arctic, primarily for subsistence. A few commercial hunts for pinnipeds continue, including those for Cape fur seals in Namibia, harp and hooded seals in Canada, and harp seals in Norway and the Russian Federation. Because pinnipeds often interact directly with fishing operations, sometimes fishermen and aqua culturists exert pressure to limit or reduce pinniped populations by commercial hunting or culling.

For manatees and dugongs, commercial hunting no longer occurs, but harvesting for meat and for dugong tusks continues in many areas. Although often illegal, there is little enforcement in most areas. The largest aboriginal hunts of dugongs persist in Australia and the western Pacific. In Australia, both Aboriginal and Torres Strait Islander people continue their traditional subsistence hunting of dugongs, but commercial hunting of dugong is prohibited.

Fisheries Interactions

While the effects of fisheries on marine mammals from entanglement and by-catch are well known, the effects of marine mammals on fish populations and the effects of fisheries on the prey base of marine mammals are less clear. The diet of most marine mammals includes fish and the possibility is often raised that some marine mammals are competing with fisheries or impeding the recovery of depleted fish stocks. Although some cetaceans, such as sperm whales, killer whales, pilot whales, and false killer whales, are known to depredate fishing operations, the significance of such depredation on fish populations and fish catches is not always clear. Similarly, the effects of pinniped predation on valuable fish populations remain uncertain. For example, in the north-eastern Pacific sea lions feed on small populations of endangered salmon during spawning migration in rivers, especially in connection with dams. On the other hand, the significance of some other interactions is less clear. Cape fur seals prey on two commercially fished species of hake off the west coast of South Africa and grey seals prey on cod populations off the east coast of Canada. Population modelling studies of these fishery interactions suggest that seal culls are likely to have small if any effects on fish catch rates. Indeed, the effect of Cape fur seal predation may be overshadowed by predation of one hake species on the other. The biological and economic significance of marine mammal interactions varies with the fisheries and fish species involved. The areas of conflict are geographically limited and fishery interactions do not appear to be a global problem.

Habitat Alterations

Disturbance

Many human activities generate underwater noise, including ship movements, military exercises involving the use of sonar or explosives, offshore oil and gas exploration and pile driving associated with construction of renewable energy facilities. Potential effects of such noise on marine mammals include direct acoustic injury, interference with foraging and communication, and food-web disruption. Disturbances are more likely to be significant for populations that have also been affected by other factors, such as harvesting or bycatch. Among the effects of disturbance that have been confirmed are the tendency of some whales to modify their movement patterns to avoid fixed-point noise sources, as for bowhead whales near an offshore oil production facility, and the mortality of deep-diving cetaceans exposed to naval sonar.

Offshore oil and gas development creates unique problems related to oil spills, whether at drill sites or during shipping. For depleted species, the additional mortality from ship strikes has been shown to be significant and vessel speed controls are demonstrably effective at reducing such strikes, as for right whales in the western North Atlantic.

Coastal and Riverine Development

Development in coastal and freshwater areas affects marine mammals in a variety of ways. Residential and urban development can make pinniped haul-out habitat inaccessible or hazardous. However, some species have shown remarkable adaptability to human-modified beach environments. For example, in the United States, California sea lions regularly haul out on piers and even on moored yachts in San Francisco. Although monk seals are extremely sensitive to disturbance in some areas, in Hawaii they sometimes rest and even nurse pups on crowded bathing beaches. In such cases, the animals seem capable of co-existing with the human presence as long as they are not molested.

Coastal habitats used by whales and dolphins make them vulnerable to increasingly intensive aquaculture operations of many types. A well-studied case is salmon farming in southern Chile, which operates in concentrated areas using open-cage net pens, moorings and anchoring, external supplementary feeding (rich in nutrients) and a significant quantity of chemical products (antimicrobials and pesticides). However, there are differences globally in fish-farm design and operation, so general conclusions are difficult.

The construction of dams, barrages, and other structures in rivers and estuaries has led to fragmentation of dolphin and manatee populations in Asia and South America, making such populations more vulnerable to various threat factors, including entrapment in canals and mortality in flood-control gates. Runoff from agricultural fields, livestock feedlots, factories, and city streets contributes to chemical and biological contamination of freshwater, estuarine and coastal food webs, with often uncertain but likely negative effects on marine mammal health. For example, sea otter deaths in California (United States), have been linked to protozoan parasites known to breed in domestic cats and toxoplasmosis has been identified in Hector's dolphins.

Climate Change

Climate change, both natural and human-induced, has the potential to affect the spatial distribution, reproductive success, foraging, and health of marine mammals. The direction of such effects, negative or positive, is likely to be variable, with some species suffering from the loss of habitat and others able to take advantage of new habitat. MacLeod predicted that the ranges of most cetacean species (88 per cent) would be affected by changes in water temperature resulting from global climate change. This author predicted that the effects would be unfavourable for about half (47 per cent) of cetacean species. Little is known about the ability of most marine mammals to adapt to rapid environmental change. For example ice seals and polar bears, which are dependent on sea ice, may be especially vulnerable to predicted climate change effects on ice habitat. Foraging habitat of right whales, which are dependent on small zooplankton, may change with increasing water temperature. Other species with more generalized diets and the ability to thrive in multiple types of habitat, such as bottlenose dolphins, may be more resilient. The overall effects of sea-level rise have been studied for northern elephant seals and Hawaiian monk seals.

Major Ecosystem Services Provided by Marine Mammals

Services to the Ecosystem

Marine mammals can affect their ecosystems in several ways. Some species, such as sea otters, dugongs and walruses, structure their foraging habitat. Depletion of these animals can result in major habitat changes; for example, kelp beds thrive when sea urchins are suppressed by sea otter predation. Other species, such as killer whales and leopard seals, play key roles as high-order predators, and their absence can affect prey resources of other marine mammal, bird or fish populations. Additionally, some species, such as sperm whales and blue whales, may have a large effect on nutrient recycling, with nutrient transport from deep ocean feeding areas to the surface. These ecosystem-level effects are understood for only a few species, but they can be critical for maintaining diverse and productive ecosystems.

Direct Services to Humans Including Economic and Livelihood Services

The economic value of products obtained from marine mammals – meat, oil, ivory, fur, and many others – has been large, and this has contributed to these animals' extreme depletion and in a few instances led to their extinction (Steller's sea cow, Caribbean monk seal). Many groups of people continue to benefit from hunting marine mammals, including in some instances from selling products in international markets (e.g., narwhal and walrus ivory and seal skins). Aboriginal people in the Arctic and sub-Arctic continue to consume products from cetaceans and pinnipeds on a regular basis. Local people in Amazonia, northern Australia, and West Africa continue to harvest sirenians for food. In Nunavut and the Northwest Territories in Canada, regulated polar bear sport hunts provide income to Inuit who serve as guides and are required to use dog teams and sleds to pursue the animals.

In contrast, many people benefit from non-consumptive or low-consumptive uses of marine mammals, especially through whale-, dolphin-, and seal-watching tourism. In addition, many people enjoy seeing marine mammals in the wild on their own. However, such activities can negatively affect small localized populations, for example bottlenose dolphins in Shark Bay, Australia.

Additionally, public display of captive marine mammals can make people more aware and appreciative of them, but it is extremely controversial, in part because the capture of marine mammals from the wild for display in captivity could threaten small wild populations.

Conservation Responses and Factors for Sustainability

Like most large animals, marine mammals have limited capacity to reproduce and increase their numbers. Therefore, factors that can result in either low recruitment (e.g. impairment of reproduction by chemical contaminants such as organochlorines in food webs; or human-induced mortality rates higher than replacement (e.g. hunting or by-catch) need to be addressed to achieve conservation goals. Harmful algal blooms, ocean acidification, and expansion of hypoxia zones are among the most intractable factors affecting marine mammal populations. Conservation requires understanding of the organisms and their habitat requirements, and a balancing of human needs and desires with the natural productivity and the carrying capacity of the environment. The identification of key limiting factors is a first step toward developing management measures that can help populations to recover.

For the most part, population recoveries are regarded as successes, although in some cases they have led to unanticipated conflicts. One such conflict is with fishermen who have become accustomed to low levels of marine mammal abundance. For example, the recovery of numbers and range of sea otters has increased competition with fisheries for high-value molluscs in Alaska (United States) and the eastern North Pacific. Similarly, as mentioned above, sea lions prey on endangered fish, such as salmon and sturgeon, and recovering grey seals in Europe have had negative effects on seabird nesting habitat. Growing pinniped populations have also led to increased interactions with recreational fishers, vessel owners, and marina managers.

The conservation of marine mammals, like conservation more generally, should be understood as a dynamic and continuing process. Consequences of management actions need to be anticipated and unforeseen consequences addressed as they arise. Especially in the case of animals like marine mammals, that is widely distributed and rarely occurs within the jurisdiction of only one State, multilateral approaches are essential. For example, the global ban on large-scale pelagic drift net fishing on the high seas imposed by the United Nations in 1994 was a major step in limiting the by-catch of several marine mammal (and seabird) species that were especially vulnerable to entanglement. Other international instruments, such as Convention on the International Trade in Endangered Species (CITES) and the International Convention for the Regulation of Whaling, have helped limit the damage caused by over-exploitation of the great whales.

With a broad understanding of the many aspects of both conservation and use of marine mammals and their roles in marine ecosystems, it should be possible to address at least some of the issues arising from human interactions.

Seabirds

Seabirds are the most threatened bird group and their status has deteriorated faster over recent decades. Globally 28 per cent are threatened (5 per cent are in the highest category of Critically

Endangered) and a further 10 per cent are Near Threatened. Of particular concern are those species whose small range or population is combined with decline (64 species). Pelagic species are disproportionately represented in comparison with coastal species; those listed under the Agreement on the Conservation of Albatross and Petrels have fared worst of all.

Declines have been caused by ten primary pressures. At sea these include: incidental bycatch (in longline, gillnet and trawl fisheries); pollution (oil spills, marine debris)); overfishing; energy production and mining. On land, invasive alien species, problematic native species (e.g. those that have become super-abundant), human disturbance, infrastructural, commercial and residential development, hunting and trapping have driven declines. Climate change and severe weather affect seabirds on land and at sea.

Given their imperilled conservation status, many seabirds have been highlighted for special conservation status and action under a range of international, regional and national agreements and mechanisms. Data on distribution, abundance, behaviour and pressures can be used to inform the design of effective management regimes for seabirds. Management decisions can be guided by: (1) where the key areas are, (2) when these areas are used, (3) what variables explain seabird presence in a given area, (4) the threat status of species in a given area, (5) what pressures may be adversely affecting the species, associated habitats and processes, (6) what management actions are needed to address these threats, and (7) how any management intervention can best be monitored to assess its effectiveness.

Seabirds provide many ecosystems services and their role as potential indicators of marine conditions are widely acknowledged. Many studies use aspects of seabird biology and ecology, especially productivity and population trends, to infer relationships with and/or effects on and/or correlate with aspects of the marine environment, particularly food availability.

Population Trends or Conservation Status Aggregated at Global Scale

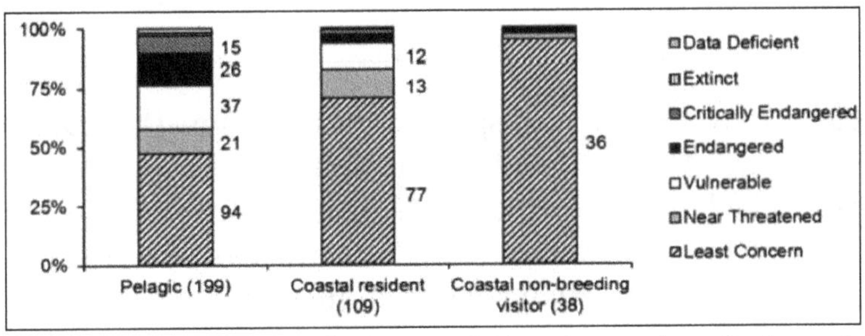

Figure: Proportion of species in each IUCN Red List category for pelagic species, coastal residents and coastal non-breeding visitors. Figures give number of species (for totals >5).

Seabirds are more threatened than other comparable groups of birds and their status has deteriorated faster over recent decades. In terms of the categories used in the International Union for the Conservation of Nature (IUCN) Red List, globally 97 species (28 per cent) are threatened, with17 species (5 per cent) in the highest category of Critically Endangered) and a further 10 per cent Near Threatened. Only four species, all storm petrels, are regarded as Data Deficient; three species are considered Extinct, and two other species are Possibly Extinct. Of the 132 threatened and Near Threatened seabird species 70 (53 per cent) qualify by virtue of their very small population and/or range. 66 species (50 per cent) qualify by virtue of having undergone rapid population decline. Of particular concern are those with both small range and/or population as well as having undergone decline (64 species; 48 per cent); this includes six species of penguins, 17 of gadfly petrels and eight of cormorants. Pelagic species are disproportionately represented in all categories in comparison with coastal species. 57 species (17 per cent) are increasing; for many, such as the 17 gull species, this is doubtless due to their abilities to exploit close links with human activities.

A broader, but less sensitive, measure of overall trends is provided by the Red List Index, which measures trends in extinction risk (based on the movement of species through IUCN Red List categories owing to genuine improvement or deterioration in status) and is virtually the only trend indicator currently available for seabirds on a worldwide and/or regional basis. It shows that, over the last 20 years, seabirds have had a substantially poorer conservation status than non-seabirds and that they have deteriorated faster over this period. Pelagic species are more threatened and have deteriorated faster than coastal species, and this difference is particularly pronounced for the albatrosses and large petrels that are covered by the 2004 Agreement on the Conservation of Albatross and Petrels.

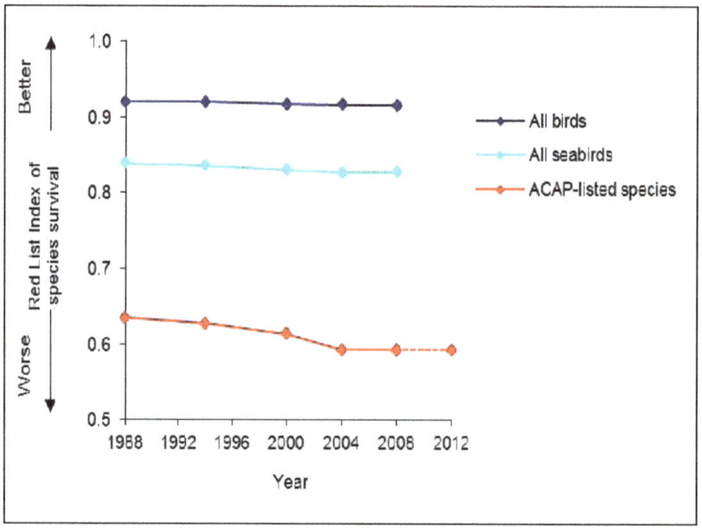

Red List Index of species survival for all bird species (n=9,853 non-Data Deficient species extant in 1988), all seabirds (n=339) and ACAP (Agreement on Conservation of Albatross and Petrels)- listed species (n=29). Values for the latter are projected to 2012 based on data from the 2012 IUCN Red List to be published later this year. RLI values relate to the proportion of species expected to remain extant in the near future without additional conservation action. An RLI value of 1.0 equates to all species being categorized as of Least Concern, and hence that none are expected to become extinct in the near future. An RLI value of zero indicates that all species have become Extinct.

Reviewing the pattern taxonomically reveals that, of the main families (which together account for 87 per cent of species), the most threatened are the albatrosses/petrels (Diomedeidae/Procellariiformes and penguins (Sphenisciformes). Together these (represent nearly one half (43 per cent) of all seabirds and contain many pelagic species. Conservation of Diomedeidae benefits considerably from ACAP. Within Procellariiformes the genera Pterodroma and Pseudobulweria are the next most threatened and a special internet forum has recently been established to promote priority conservation action for them.

For Major Taxonomic and/or Geographic Subdivisions

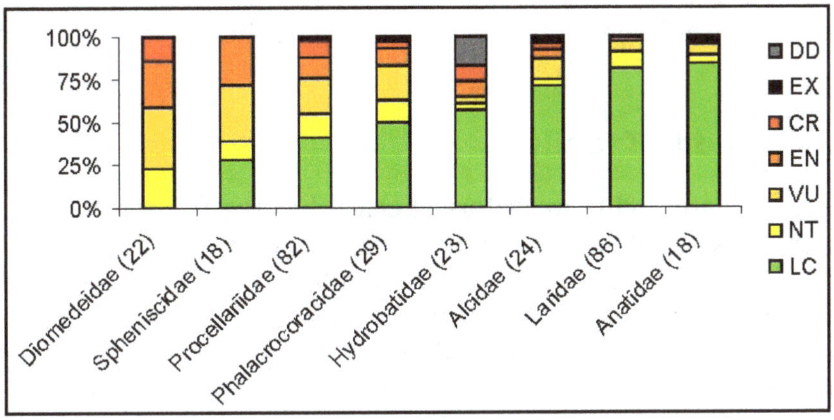

Figure: Percentage of species in each IUCN Red List category for the major seabird families. Figures give number of species.

Special Conservation Status Issues

Given their imperilled conservation status, many seabirds have been highlighted for special conservation status and action under a range of international, regional and national agreements and mechanisms. However, because seabirds are highly mobile and migrate, they are exposed to vagaries of differing levels of protection across international (and non-governmental) regions. Those agreements and mechanisms currently most actively undertaking work include ACAP (30 species), EU Birds Directive (all seabirds in the EU), the Convention for the Protection of the Marine Environment of the North-East Atlantic3 (OSPAR Convention) (9 species), the Agreement on the Conservation of African-Eurasian Migratory Water birds (82 species), East Asian-Australasian Flyway Partnership (39 species), the Convention for the Protection of the Mediterranean Sea Against Pollution5 (Barcelona Convention) (14 species), Convention on the Conservation of Migratory Species of Wild Animals (CMS; 20 seabird species are listed on Annex I; 50 on Annex II), the Convention on the conservation of European wildlife and natural habitats7 (Bern Convention) (over 30 species), Helsinki Commission (HELCOM; 11 species), the Convention on the Protection of the Black Sea Against Pollution8 (Bucharest Convention) (2 species), Commission for the Conservation of Antarctic Marine Living Resources (CCAMLR; 7 species), Convention for Arctic Flora and Fauna (3 species), Migratory Bird Treaty Act (139 species), North American Agreement on Environmental Cooperation (1 species), Trilateral Committee for Wildlife and Ecosystem Conservation and Management (1 species), and the Convention on International Trade in Endangered Species of Wild Fauna and Flora (6 species). Other agreements that have this remit but are not yet active include the Nairobi Convention for the Protection, Management and Development of the Marine and Coastal Environment of the Eastern African Region (Nairobi Convention) (47 species), the Regional Convention for the Conservation of the Red Sea and Gulf of Aden Environment (Jeddah Convention) (lists not yet provided by contracting parties), the Convention for Cooperation in the Protection, Management and Development of the Marine and Coastal Environment of the Atlantic Coast of the West, Central and Southern Africa Region (Abidjan Convention) (considering adding a species list), and the Convention for the Protection and Development of the Marine Environment in the Wider Caribbean Region (WCR) (Cartagena Convention) (5 species). In addition to the above agreements, Regional Fisheries Management Organisations (RFMOs) have also begun to adopt strategies that address incidental seabird bycatch. Level of regulation varies across RFMOs but includes combinations of the use of one or more bycatch mitigation measures in certain areas, data collection through observer programmes and use of monitoring, surveillance and compliance measures.

Key Pressures Linked to Trends

The majority of seabirds are highly migratory species that require a variety of marine and terrestrial habitats during different seasons and life stages. Many seabirds are long-lived and slow reproducing. These characteristics make them particularly vulnerable to a wide range of pressures, where even quite small increases in mortality can lead to significant population declines. In addition, many seabirds have highly specialised diets, being reliant on just a few prey species, the abundance and distribution of which can alter dramatically in response to abrupt environmental changes.

Croxall et al. found that globally, of the top 10 pressures on threatened seabirds, invasive species typically acting at the breeding site potentially affect 73 species (75 per cent) of all threatened seabird species and nearly twice as many as any other single threat, although in some cases the threat is of a potential future impact. The remaining pressures are fairly evenly divided between: (a) those acting mainly at the breeding site, namely problematic native species (e.g. those that have become superabundant - 31 species, 32 per cent), human disturbance (26 species, 27 per cent), infrastructural, commercial, and residential development (14 species, 14 per cent) and (b) those acting mainly at sea in relation to foraging, moulting or migration areas/aggregations, namely, bycatch in longline, gillnet and trawl fisheries (40 species, 41 per cent), pollution (30 species, 31 per cent), overfishing and/or inappropriate spatial management of fisheries (10 species, 10 per cent). Hunting and trapping (23 species, 24 per cent) and energy production and mining (10 species, 10 per cent) affect both domains, the former more at breeding sites, the latter more in relation to foraging areas, flight paths and flyways. Climate change and severe weather (39 species, 40 per cent), as currently assessed, largely reflect adverse weather and flooding at breeding sites. However, the impact of sea level rise is clearly an important driver of change that is increasingly affecting seabirds in many ways, albeit mainly in the medium to long term (i.e., at time frames mostly outside those of relevance to IUCN Red List criteria). The relative importance of threats is largely similar when only those of high impact are considered, although bycatch becomes almost as significant as the effects of invasive alien species.

Commercial fisheries are the most serious at-sea pressure facing the world's seabirds, affecting both adult and juvenile birds. Despite data gaps, each year incidental bycatch in longline fisheries is estimated to kill 160,000-320,000 seabirds from 70 species, although there is evidence of substantially reduced bycatch in some key fisheries where the pressure has been managed. Several papers have reviewed seabird bycatch rates in both demersal (bottom) and pelagic (upper water column) longline fisheries in various regions, and two assessments have been made on a global scale. The fleets identified as having the highest levels of seabird bycatch include the Spanish hake fleet in the Gran Sol area, the Japanese pelagic tuna fleet in the North Pacific, the Namibian hake fleet and the Nordic demersal fleet. The impacts of illegal, unreported, and unregulated fishing (IUU) on seabirds have been estimated in the thousands of individuals each year south of 30° S but are inherently difficult to assess here and elsewhere.

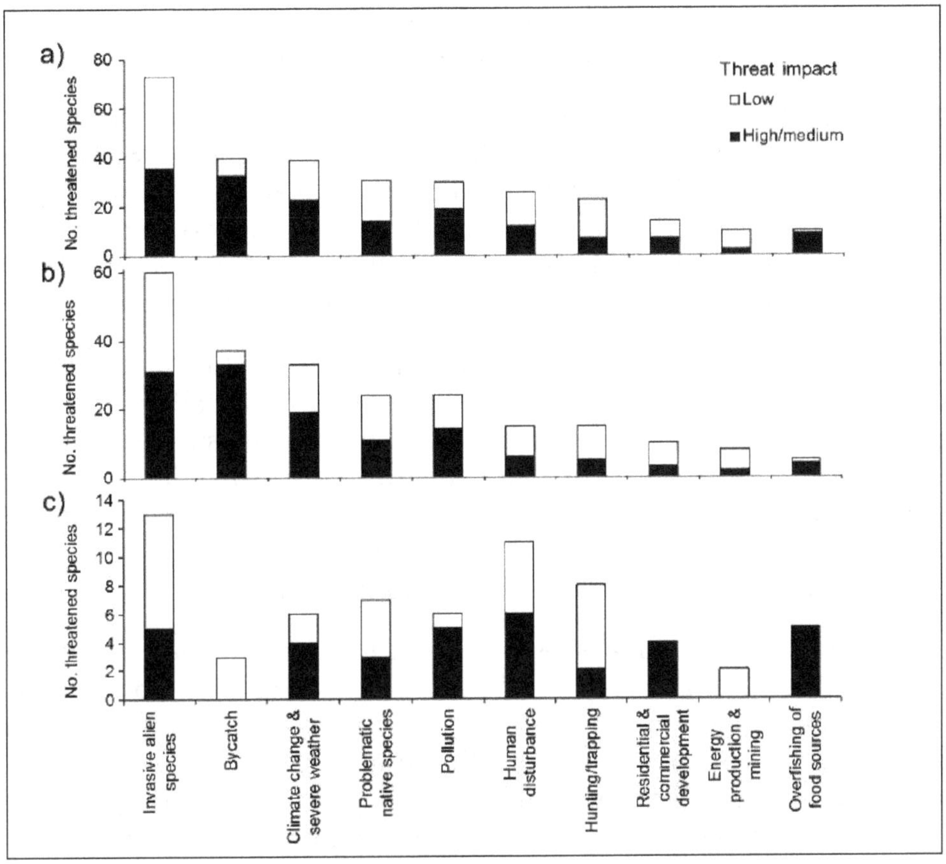

Figure: Threats to threatened (a) seabirds (n=346 species); (b) pelagic seabirds (n=197 species); (c) coastal seabirds (n=146 species).

Since 1992 a global moratorium has been imposed on the use of all large-scale pelagic drift-net fishing on the high seas, including enclosed and semi-enclosed seas (General Assembly resolution 46/215). Gillnet fisheries (both set and drift nets) are, however, still permitted to operate within a State's Exclusive Economic Zone (EEZ). Although many data gaps remain, hampering assessment, a review of existing data shows that gillnets are responsible for the incidental capture of large numbers of birds, sharks and marine mammals. Amongst birds, the pursuit-diving species, such as divers (loons), grebes, sea ducks, auks and cormorants, are the most vulnerable to entanglement. The most recent global review estimated incidental bycatch in gillnet fisheries at 400,000 seabirds from 150 coastal and diving species each year. The highest bycatch has been reported in the North-west Pacific, Iceland and the Baltic Sea.

Although seabird bycatch in long-line fishing has been known since the 1980s, the threat posed by trawl fisheries has also become apparent in recent years. No global review of the impact of trawl fishing on seabirds has been undertaken, but a number of regional and national levels studies highlight the significance of the problem with tens of thousands of 40 larger species of seabird thought to be killed each year. Trawling can also alter benthic habitats which may have indirect impacts on seabirds via the effect this has on forage fish species.

Fisheries may compete with seabirds for their prey items, and overfishing of both forage species and predatory species that help aggregate food sources for seabirds have been cited as a reason for the decline in several species. Cury et al. assessed prey abundance and breeding success for 14 bird species within the Atlantic, Pacific, and Southern Oceans and found that when less than one third of the maximum prey biomass was available to seabirds, their productivity was adversely affected.

Climate change and severe weather driven by habitat shifts and alterations, storms and flooding, and temperature extremes are already affecting some seabird species. Species' sensitivity and adaptive capacity depend on a suite of taxon-specific biological and ecological traits; as well as the degree to which they are exposed to changes in climate. Known negative impacts may include loss of habitat, decreased marine productivity causing shifts in location of prey, and shifts in range and migration routes due to changes in winds, ocean currents and sea surface temperature.

Pollution in various forms is a widespread problem adversely affecting many seabirds. Oil spills, from both offshore facilities and shipping tankers, can cause mortalities that lead to population-level impacts, particularly when they occur within the most sensitive sites. Single spills have been recorded as killing up to a quarter of a million birds and causing the loss of 7 per cent of regional populations of certain species. Since its advent, plastic in the form of solid waste materials has become ubiquitous in all oceans of the world and entanglement and ingestion of this material by seabirds is now a widespread problem, affecting at least 100 species.

Attraction to artificial sources of light has been recorded in at least 21 species of Procellariiformes, as well as in several other seabird groups, and has a detrimental effect on some globally threatened populations, notably shearwater and Pterodroma species around their breeding colonies. Light-induced seabird collisions at sea, either with fishing vessels (such as those emitting light to catch squid) or with marine oil platforms, are difficult to quantify, occurring episodically particularly in low-visibility conditions and probably exacerbated by seabirds' attraction to bright lights and flares. However, up to tens of thousands of seabirds have been observed in a single collision event.

Impacts from shipping may include water and air pollution, disturbance, and collision. The level of impact may increase in the future as ship traffic increases, particularly in sensitive areas such as the Arctic, where key seabird habitats and potential shipping routes may overlap, and further exacerbate impacts from predicted climate change.

Major Ecosystem Services Provided by the Species Group and Impacts Services to Ecosystems

The role of seabirds as potential indicators of marine conditions is widely acknowledged. Many studies use aspects of seabird biology and ecology, especially productivity and population trends, to infer relationships with and/or effects on and/or correlate with aspects of the marine environment, particularly food availability.

Seabirds play a key role in nutrient cycling via the shaping of the plant community in their terrestrial and coastal breeding habitat. Seabirds transport allochthonous nutrients (i.e., fixed nitrogen, phosphorus, and trace elements), mainly via their guano, to seabird colonies (i.e., cross-ecosystem subsidies). They also shape plant communities in their breeding habitat by creating physical disturbance, dispersing seeds, and bioturbating the soil with their burrowing. These functions provided by seabirds increase productivity and diversity in terrestrial and coastal ecosystems surrounding seabird colonies.

Direct Services to Humans Including Economic and Livelihood Services

Seabirds contribute several provisioning (e.g., protein, guano) services, play an important cultural role in many countries (e.g., for the Maori of New Zealand and the Tsimshian of Alaska), and feature in Greek, Hawaiian and Christian mythology. Seabird breeding colonies are increasingly used as a means to generate tourism income.

Seabird guano has excellent properties as a natural fertilizer enriching both terrestrial and marine environments. It consists of nitrogen-rich ammonium oxalate and urate, phosphates, as well as some earth salts and impurities. It typically contains 8 to 16 per cent nitrogen (the majority of which is uric acid), 8 to 12 per cent equivalent phosphoric acid, and 2 to 3 per cent equivalent potash. Archaeological evidence suggests that Andean peoples have collected seabird guano for well over 1,500 years. A harvest boom in the nineteenth century, called the "white gold rush", saw tens of thousands of workers extracting guano from the Peruvian seabird breeding islands and loading thousands of tons onto each ship. Other harvested guano islands were located in the Caribbean, atolls in the Central Pacific, and off the coast of Namibia, South Africa, Oman, Patagonia, and Baja California. This unsustainable harvest resulted in massive deposits of guano, in some cases more than 50 m deep, being severely depleted. Many areas of the industry collapsed, although some Peruvian islands are still managed for guano on a rotational system.

Harvesting of seabird adults, chicks, eggs, and feathers have been important activities for some coastal communities for many centuries, but have also driven seabird declines. Bones were used to make fishing hooks and musical instruments and to engrave tattoos; feathers featured prominently in the millinery trade and are still used in some countries for local arts and handicrafts, e.g., to make cloaks and hair adornments. The meat and eggs still form key sources of protein. Harvest methods have changed over time to include more efficient tools, making the seabirds more exposed to excessive harvesting. Declines in a number of species have been attributed to over-exploitation. Harvesting quotas exist in some areas, such as in the Seychelles (limited to 20 per cent of Sooty Tern eggs each year), New Zealand (limited to 13 per cent of Sooty Shearwater chicks each year, and the United Kingdom (limited to 2,000 Northern Gannet each year).

However, unregulated harvesting is a substantial problem in the entire Arctic region (2 million adults and countless eggs of several species of Alcidae are taken each year), the Tuamotus and the Marquesas (egg collection), Peru (Waved Albatross and Humboldt Penguin), Madagascar (egg collection), Jamaica (egg Collection) and Indonesia.

For centuries fishers have used seabirds as a visual guide to locate fishing areas. They remain important for artisanal operations (such as in Hawaii, Comoros, Madagascar and Tanzania), which search for flocks of seabirds in order to find fish. Without seabirds, these livelihoods (e.g., catching small skipjack and juvenile yellow-fin tuna) could disappear or be substantially adversely affected.

Viewing seabirds is an increasingly popular pastime for many tourists; many spectacular breeding colonies are accessible to visitors and revenues generated contribute substantially to local economies. For example, in Australia, the Phillip Island Little Penguin colony receives half a million visitors a year, spending 35 million Australian dollars. A single African Penguin colony in South Africa generates United States dollars 2 million/yr in tourist revenue. In New Zealand, nature-based tourism relying primarily on the Yellow-eyed Penguin returned 100 million dollars annually to the Dunedin economy; hence a single breeding pair could be worth 60,000 dollars/yr. The Royal

Society for the Protection of Birds (RSPB) estimated that four of its seabird reserves in the UK (one each in England, Northern Ireland, Scotland and Wales) together generated around 1.5million dollars/yr for the local economies. Tourism in the Galapagos is thought to generate over 62 million dollars each year; seabirds are a prime reason for visiting. Pelagic trips to view seabirds at sea have also become popular, particularly in Europe, North America and the Southern Ocean. The value of these trips has not been quantified to any degree, but is likely to be significant; for example, 80,000 dollars was spent on a single pelagic trip off South Africa.

Conservation Responses and Factors for Sustainability

Data on seabird distribution, abundance, behaviour and pressures can be used to inform the design of effective management regimes. Management decisions can be guided by: (1) where the key areas are, (2) when these areas are used, (3) what variables explain seabird presence in a given area, (4) the threat status of species in a given area, (5) what pressures may be adversely affecting the species, associated habitats and processes, (6) what management actions are needed to address these threats, and (7) how any management intervention can best be monitored to assess its effectiveness.

Depending on the species, the priority actions needed may involve: (a) formal and effective protection of the most important sites. For site protection to be effective, it should ensure that areas are large enough to capture critical behaviour (such as key breeding sites, the marine areas around them used for maintenance and more distant feeding and aggregation sites), consider temporal and spatial variations, and have adequate regulation to minimise effects of any pressures. Where national, regional and global networks of Marine Protected Areas (MPAs) are being developed, inclusion of key sites in those networks would contribute substantially to the necessary site protection; (b) removal or control of invasive, and especially predatory, alien species from areas used for seabird breeding, feeding and/or aggregation, as part of habitat and species recovery initiatives; and (c) reduction of bycatch to levels that do not pose a threat of species decline. For many uncommon species or species of low productivity, this likely can only be achieved when bycatch is reduced to near zero. Other, more generic actions, such as education and awareness-raising and accompanying stakeholder involvement, are also high priorities, as are some more species-specific activities, such as harvest management, species reintroductions and species recovery. Although it is relatively straightforward to derive these generic recommendations for conservation action, it can be costly and difficult to implement them effectively and at a sufficient scale to make a difference to the conservation status of seabird species. However, progress has been achieved in recent years in terms of the three highest priority actions, but despite these successes, problems will continue without further action.

Where simple seabird mitigation measures have been implemented, there is evidence of substantially reduced bycatch in some key fisheries where the pressure has been managed, including a greater than 95 per cent reduction in some areas. The main tuna RFMOs now have voluntary or binding regulations in place that require the use of a combination of mitigation techniques in different geographies, though their effectiveness may be hampered by a lack of monitoring and/or enforcement.

Key sites for seabirds have begun to be protected in several countries, primarily covering selected breeding sites on land, though marine designation for seabirds has also advanced, with new MPAs in Europe, the Antarctic and the Americas in recent years. Where eradications and/or controls of invasive alien species have been undertaken, recoveries of seabird populations have been rapid and dramatic, and a great number of larger islands are now being tackled. Translocations of some species to new locations have also proved an effective conservation strategy for several species.

Actions that are implementing an ecosystem approach to capture fisheries management are discussed in Part of this assessment; many of those measures, including better management of selectivity of fishing gear and including ecosystem feeding requirements in setting fishery harvest limits, will contribute to improving the conservation status of seabirds if implemented effectively.

Amphibians

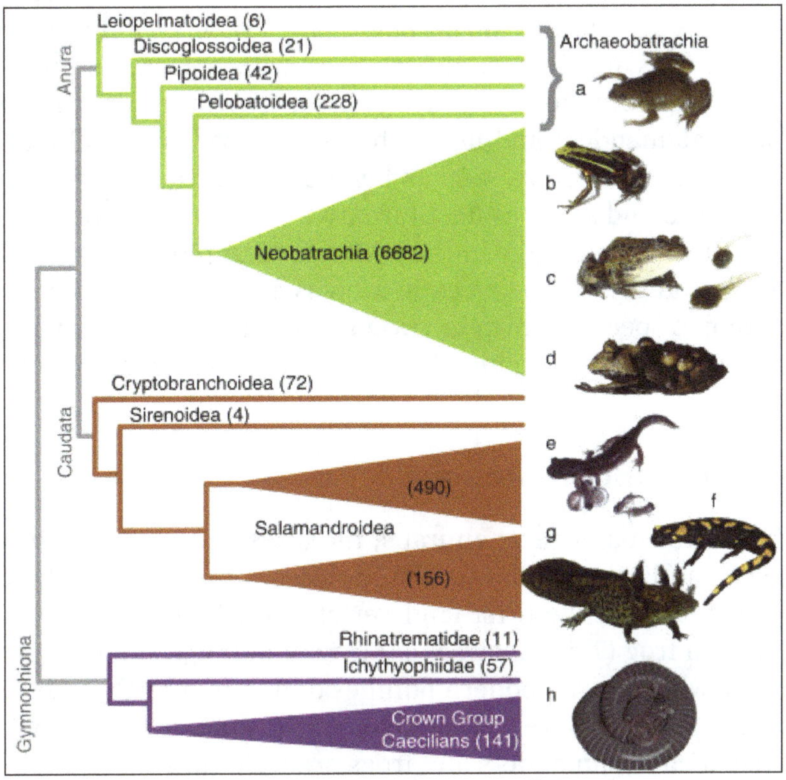

Figure: The amphibian tree of life.

The three orders of amphibians differ greatly in numbers, morphology, and reproductive modes. Generalized phylogeny is shown with the number of species for major clades in parentheses. Amphibians shown include: (a) Xenopus leavis (Pipidae), a fully aquatic species model organism; (b) Epipedobates tricolor (Dendrobatidae), a toxic species that sequesters toxins from its diet; (c) Rana aurora (Ranidae) has a 'typical' frog life history with eggs laid in a pond where they hatch and grow as tadpoles until metamorphosis into an adult form; (d) females of the frog Hemiphractus johnsoni (Hemiphractidae) carry developing eggs on their backs, where they hatch into fully formed miniatures of adults; (e) Ensatina eschscholtzii (Plethodontidae) females guard their eggs, which hatch into fully formed miniatures of adults; (f) Salamandra salamandra (Salamandridae), the Fire Salamander of European folklore; (g) Ambystoma andersonii (Ambystomidae), a critically endangered permanently larval form closely related to the model organism

A. mexicanum, the Axolotl; (h) Siphonops annulatus (Siphonopidae), a caecilian whose females stay with their eggs until they hatch into miniatures of adults and feed on a lipid secretion produced by the skin.

Amphibians have been a part of human culture. Western Europeans built fi res for cooking and warmth, adding large logs as needed. What occasionally emerged was astounding: large black animals (which had found shelter in the logs) with four legs and a tail, jet black with striking bright yellow spots. These fire salamanders were variously thought to be the product of the fire itself, or, as Aristotle reported, capable of extinguishing fi re. Pliny the Elder is said to have tested this idea by throwing a salamander into flames — the salamander died! — Nevertheless the association with fire persisted. Pliny perpetuated other fantastical claims, which spread; even Leonardo da Vinci contributed to the legend, and myths from different regions merged — at one point, asbestos was claimed to be salamander wool. Salamanders were attributed great powers; a single salamander upstream was thought to be sufficient to kill an army. King Francis I. of France chose a salamander as his emblem — a powerful symbol, born of fi re, filled with poison, immune from burning, and even able to douse flames. Before the emergence of great cities and conurbations, people grew up surrounded by nature. Salamanders and newts, toads and frogs were all part of normal human experience. In general, amphibians are seen as benign and harmless, even helpful as creatures that devour harmful insects and serve as an alternative food source. Thus, it came as a shock to most biologists and to the public at large in the 1980s that amphibians around the world were in decline and that they were at greater risk of extinction as a taxon than any other vertebrate group. A study of every amphibian species known in 2004 showed that on the order of 40% were at high risk of extinction, and by 2008, the decline of amphibians was seen as evidence of an impending sixth mass extinction.

Amphibian Taxonomy and Anatomy

Amphibians constitute three major taxa: anurans, the frogs and toads; caudates or urodeles, the salamanders and newts; and gymnophiones, the limbless caecilians. Amphibians are an ancient group, representing remnants of the fi rst land vertebrates. Modern taxa are early Mesozoic in origin. The earliest known frog (Triadobatrachus) lived about 250 million years ago (early Triassic). They are survivors, with several modern families dating to mid-Cretaceous (110–120 million years ago); amphibians appear to have passed through the mass extinction at the end of the Cretaceous relatively unscathed. Salamanders and frogs are well-known and broadly distributed, but caecilians remain poorly known, even to professionals. Caecilians are mainly burrowers (fossorial) that rarely are seen on the surface; little is known of their life history. Today nearly 8,000 living amphibian species are known, of which 88% (~7000) are frogs (anurans), about 9% (~700) are salamanders, and about 3% (>200) are caecilians. Frogs and salamanders are more closely related to each other than either is to caecilians. These three groups are distributed very differently around the world. Frogs are found on all continents except Antarctica and on most large and small continental islands in the temperate and tropical regions (for example, they are abundant on Madagascar, the Philippines, Borneo and New Guinea), but absent from all but a very few oceanic islands. They are especially abundant in the tropics. In contrast, salamanders are distributed mainly in northern regions (although they reach below the equator in South America). Salamanders are absent even from most large continental islands and all oceanic islands. Most restricted of all, caecilians are limited to the tropics.

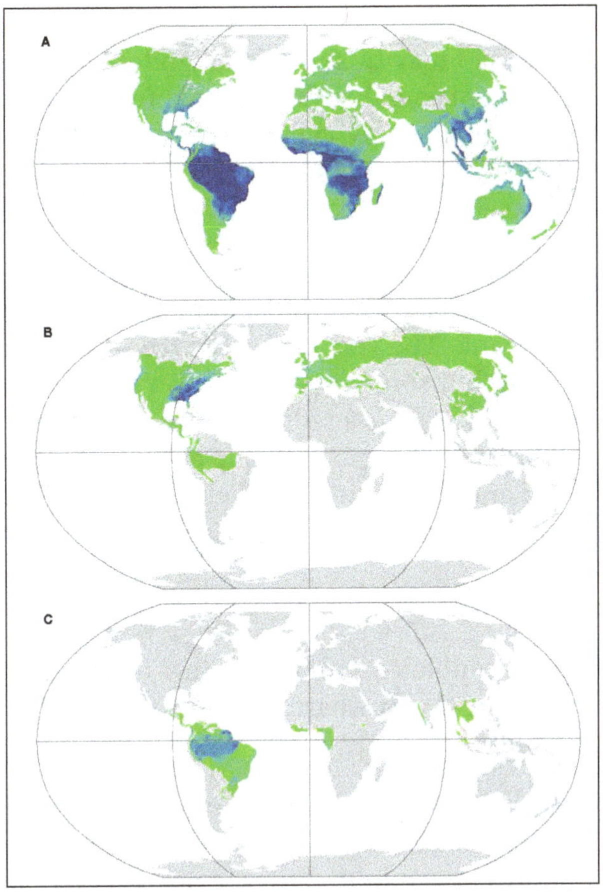

Figure: Global distribution of the three orders of amphibians.

Figure shows The three orders of amphibians differ in distribution with frogs (A) commoner to the south, salamanders (B) to the north, and caecilians (C) restricted to the tropics. Darker colors toward blue indicate higher species richness.

The three groups of amphibians are so distinct in appearance that they never are mistaken for one another. What they most notably share is a moist skin, which serves as the main respiratory organ and produces diverse secretions, many of them poisonous to other organisms. The frog body plan is highly conserved and they are unmistakable for any other vertebrate. Frogs have very short trunks, with nine or fewer vertebrae in front of the sacrum. Their bodies are made larger by the elongated pelvic girdle. All adult frogs lack tails. Forelimbs are relatively short, but hind limbs are long. Frogs have four digits on the forelimb and five on the hind limb, but sometimes one or two are very small. The hind limbs contain four well-defined segments and can be very long, enabling some species to make prodigious leaps. Leaps exceeding two meters have been recorded. However, many species that live mainly underground have short legs and walk or hop, or burrow, rather than leap. Heads of frogs are large relative to body size.

Salamanders typically have the body type of a prototypical, generalized land vertebrate, with well-proportioned heads, limbs of roughly similar length, and a tail about as long as the head plus body. Trunks are of moderate length with between 12 and 22 presacral vertebrae and the tail is sometimes much longer then the body. Limbs are of moderate size and usually overlap when pressed against the trunk; the limbs have three major segments. Salamanders are like frogs in having no more than four digits on the forelimb, while five typically are found on the hind limb. Because of their appearance, people may mistake salamanders for lizards and vice versa. A few exceptions include aquatic amphiumas and sirens, salamanders that have become elongate and eel-like with reduced or absent limbs. Aptly called giant salamanders of China and Japan can exceed 1.5 meters in length. Other salamanders have retained aquatic larval morphology, usually with feathery external gills and keeled tails, throughout life.

Caecilians differ sharply from other amphibians. They are superficially worm-like, with bodies marked by rings. Most are terrestrial burrowers, but a few are aquatic. The head is small and much the same width as the anterior trunk. The very elongated trunk often contains well over 100 vertebrae. The tail is typically absent, but some species have a very short tail. The cloacal opening is usually at the very end of the animal. Eyes are reduced to nearly but never completely absent. The mouth is large with many teeth. The head has a unique chemosensory organ, the tentacle, which lies between the external nostril and the mouth and flicks in and out like a snake's tongue. Limbs or limb rudiments are completely absent, even during development.

Amphibian Biology

Water governs the lives of all amphibians, although a few frogs and salamanders have gone to extremes of specialization to live in places where water is scarce. Many frogs and especially the tropical salamanders live on trees.

What is often presented as a typical amphibian life history is strongly biased by the fact that early biologists who first described amphibians resided in temperate zones, where the 'typical' amphibian life history dominates: land-dwelling adults gather at ponds or streams in spring to meet, mate and reproduce. Females lay eggs in water, which are then fertilized by males. Tadpoles (in the case of frogs) or larvae (salamanders) are produced that grow and metamorphose into tiny juveniles, which return to land. Europeans knew about some alternative life histories; the midwife toads (Alytes) lay a few large eggs in water, which are collected by males and draped around their hind limbs until they are ready to hatch, when they were returned to ponds.

Thousands of species of frogs, hundreds of salamanders, and nearly all caecilians diverge from the typical life history. The most common difference is loss of the larval stage; eggs that develop directly into miniatures of the adults are laid on land. A stark contrast is found in some salamanders that remain in a larval state (usually with gills and fins) throughout life, a condition often termed paedomorphosis or neoteny. In many caecilians, eggs are retained in modified oviducts and nourished before — sometimes very large — miniatures of adults emerge; this is true viviparity. Both frogs and salamanders have a few viviparous species that give birth to metamorphosed young that have been nourished, but they are uncommon. Instead, especially in frogs, almost endless variations are found on the theme: stay out of water, where danger lurks. Eggs are laid in bromeliads on tree branches, placed in vocal sacs of males, in compartments on the backs of females, even in the stomach! Larvae in bromeliads are fed nutritive eggs, or they have many specializations for living in unusual places. Reproductive diversity is one of the hallmarks of amphibians. Their skin distinguishes amphibians from other vertebrates. It is well vascularized and serves as the main respiratory organ. It is richly supplied with diverse glands that produce an amazing array of secretions, some of which help moisten the skin while others produce a broad array of poisons and toxins. Chromatophores in the skin produce the often dramatic colors that characterize many species.

Figure: Amphibian warning colors.

Aposematic (warning) coloration has evolved many times in amphibians, and mimicry is widespread. Top: these poison dart frogs live together in Ecuadorian and northern Peruvian rainforests: Ameerega bilinguis (left) is toxic and distasteful to predators, which the harmless Batesian mimic Allobates zaparo (right) uses to its advantage. Bottom: the polymorphic salamander Ensatina eschscholtzii (right; Plethodontidae) mimics the highly toxic Rough-skinned newt Taricha granulos (left; Salamandridae) in parts of its range where the two species overlap, and only here does it closely resemble the newt. Pictured specimens are displaying their bright eye and ventral coloration.

Several amphibians qualify as 'model organisms', which are used in experimental biology. The most famous are the Axolotl (Ambystoma mexicanum; especially important in studies of limb and tail regeneration) and the tetraploid African clawed frog (Xenopus laevis) and its diploid relative X. tropicalis, both widely used in studies of developmental and cell biology. A few additional frog and salamander species are widely used, including the Iberian ribbed newt (Pleurodeles waltl). There are two major impediments with amphibians as model organisms: husbandry (which explains why there is no caecilian model organism as there is no captive breeding protocol) and lack of complete sequenced genomes. Genomes of the two Xenopus have been available for a few years, but two salamander genomes have become available only now. Few other complete genomes have been obtained because the very large genomes of many frogs and all salamanders are a major impediment. Genome sizes range as high as nearly 15 Gb for frogs and well over 100 Gb for salamanders. Species with very large genomes experience greatly reduced cell cycle times; slow development and long generation times, as well as very low basal metabolic rates. These factors would seem to be highly detrimental and one would expect natural selection to have weeded out the extraordinarily abundant and highly repetitive DNA was it not for some compensating factor. The nature of that compensation, however, remains unknown and constitutes a major research challenge because for the species with the largest genomes the costs are great (e.g., increased developmental time, extended age at first reproduction).

Threats to Amphibians

The first worldwide assessment of the status of amphibians by the International Union for Conservation of Nature (IUCN) published in 2004 found an astounding 43% of species to be experiencing population decreases; 7.4% were in the highest category of risk. This is the highest level of endangerment of any major vertebrate taxon and is especially worrisome because 22.5% of the then-known species (5743) were so poorly known that they could not be assessed. These species are also likely to be in high-risk categories. In the intervening years, the situation has only worsened and the proportion of insufficiently known species has increased.

While amphibians might appear to be vulnerable to environmental challenges because of their moist skins and life history traits, they are, in fact, examples of resilience, having survived previous mass extinction events and being successful in diverse habitats from lowland rainforests to deserts and extreme high elevations. Their resilience makes it all the more challenging to explain some of the current declines.

Figure: Amphibian threats.

Amphibians are being hit hard by pathogenic fungi. Top: the blue-tailed fi rebellied newt (left; Cynops cyanurus) and oriental firebellied toad (right; Bombina orientalis) are Asian species suspected to be agents in the spread of chytrid fungus worldwide, in part due to their popularity in the pet trade. Bottom: the harlequin frog (left; Atelopus varius) from Central America through Colombia, and salamandra chica (right; Bolitoglossa chica) from northwestern Ecuador have experienced precipitous declines attributed mainly to chytridiomycosis outbreaks.

The known causes for amphibian declines are many and complex and include such widely understood factors as habitat destruction, the clearing of forests for timber and replacement with agricultural land, expansion of urbanization and drainage of wetlands. Other factors include the widespread use of pesticides and fertilizers and the influence of increasing environmental temperatures and droughts associated with climate change. What has attracted the greatest attention, however, is a pathogenic chytrid fungus of very wide distribution that has had a demonstrated devastating impact on amphibians worldwide. The pathogen, Batrachochytrium dendrobatidis (or Bd), was discovered only at the end of the last century and since then has become a major focus of investigation. The microscopic organism invades the permeable skin and disrupts the osmotic balance of the infected individual, ultimately killing it by inducing heart failure. Bd was first detected in outbreaks of extreme mortality in amphibian-rich environments of the Australian wet tropics and upland cloud forests in Costa Rica and Panama. Massive mortality also occurred in formerly abundant mountain yellow-legged frogs in the high Sierra Nevada in the iconic national parks (Yosemite, Sequoia Kings Canyon) of California. Subsequently, outbreaks have been traced to the early 1970s in Mexico, and there is circumstantial evidence of even earlier events elsewhere. The fungus is likely to have arisen in eastern Asia, where it is endemic and seems to cause little harm, and to have spread out, possibly from Korea, near the middle of the 20th century. Mitigation of the effects of the chytrid has included promising studies of the skin microbial community to enhance the frog's own immune system. In some areas, Australia for example, there has been recovery from the initial outbreaks. At present, however, Bd constitutes the major threat to large numbers of frog species and communities.

Returning to the Fire Salamander of European legend, it is deeply ironic that being the epitome of a survivor, even of fire, they are suffering a steep decline in one of their last redoubts in The Netherlands, a small forested area just north of Maastricht. In this area, researchers recently discovered another new, highly virulent, single-celled chytrid fungus, Batrachochytrium salamandrivorans (Bsal). In addition to the chaos caused by Bd, an even more deadly pathogen is now wreaking havoc. Fortunately, the new chytrid fungus is largely restricted to the immediate site of discovery, but in the time since discovery, it has been found in nearby parts of Belgium and Germany and the threat of spread to other species and areas is high. It seems to be endemic and enzootic in species of the Family Salamandridae in eastern Asia and it is likely to have been introduced to Europe through the pet trade. In both Europe and North America, new restrictions have brought importation nearly to a halt. Whereas Bd can kill individual amphibians, some recover, and some immunity has been detected, so far as is known, outside its native range, Bsal is lethal. Other infectious diseases also affect amphibians. While most are relatively poorly known and not associated with any mass mortality events, ranaviruses have been known for decades and they can be a significant factor in local populations.

Amphibian Diversity and Distribution

Paradoxically, in this era of great concern for amphibian survival, more and more new species are being discovered. Some of these are not new discoveries but rather subdivision of long-known species detected through the use of molecular markers. The majority are new discoveries. The numbers are impressive. As of August 2018, 7900 species of amphibians are known, about twice the species known in 1985. Roughly three new species are being described per week! Many of the new species reflect the greatly increased focus on the study of amphibians and the numbers of scientists being recruited. Field biologists are probing ever deeper, into the last unexplored places on this planet. Areas that are particularly rich in new species include Madagascar, Papua New Guinea and several South American countries. It is not just unexplored areas that are producing new species; in 2017 and 2018, new species were recorded from the United States and many new species were found in urban areas of India and China.

South America is the richest area for frogs, with Brazil having over 1000; Ecuador has 586 frogs, which contrasts with 28 for Spain, a country nearly twice as large. Salamanders are more northern in distribution, with United States having almost 200 species. While salamanders in the Old World barely enter the tropics, the lungless salamanders, Plethodontidae, have undergone an adaptive radiation in the New World tropics (over 40% of all salamanders), where they extend south of the 20th parallel in Bolivia. Caecilians occur only in tropical countries. Perhaps the most surprising center of caecilian diversity is the Seychelles islands (with eight species) in the Indian Ocean; India has the most caecilians (nearly 40 species).

Amphibians are key components of ecosystems throughout the world, and they may be the most abundant vertebrate in many local communities despite escaping notice by most people. They are often the top predator of invertebrates in ecosystems, and the most abundant vertebrates by biomass. Frogs and salamanders are a critical food source for other vertebrates. They are richly represented in tropical regions where they have radiated especially along mountain slopes. Frogs and salamanders are familiar animals, enjoyed by the public, studied by many scientists, and the focus of biomedical and environmental research. Today amphibians are in crisis, often victims of anthropogenic factors and novel pathogens, but they are long-term survivors and many species continue to thrive. Frogs, salamanders and caecilians promise to be with us for the long term.

Chapter 3

Marine Invertebrates

Marine invertebrates are animals that neither possess nor develop a vertebral column, and live in marine habitats. It includes all types of marine sponges, cnidarians, worms, echinoderms, molluscs and arthropods. This chapter closely examines these marine invertebrates to provide an extensive understanding of the subject matter.

Marine Sponges

Sponges have been around for a very long time, with certain species having a fossil record that dates back approximately 600 million years to the earliest (Precambrian) period of Earth's history. The approximately 8,550 living sponge species are scientifically classified in the phylum Porifera, which is comprised of four distinct classes: the Demospongiae (the most diverse, containing 90 percent of all living sponges), Hexactinellida (the rare glass sponges), Calcarea (calcareous sponges), and Homoscleromorpha (the rarest and simplest class, only recently recognized, with approximately 117 species).

Figure: A variety of sponges dot the seascape of Flower Garden Banks National Marine Sanctuary. From round to encrusting to branching, the colors and textures of sponges add to the complexity of the reef.

While sponges, like corals, are immobile aquatic invertebrates, they are otherwise completely different organisms with distinct anatomy, feeding methods, and reproductive processes. The main differences are:

- Corals are complex, many-celled organisms. Sponges are very simple creatures with no tissues.
- All corals require saltwater to survive. While most sponges are found in the ocean, numerous species are also found in fresh water and estuaries.

Regardless of these differences, sponges are important inhabitants of coral reef ecosystems. A diverse sponge population can affect water quality on the reef as the sponges filter water, collect bacteria, and process carbon, nitrogen, and phosphorus. In nutrient-depleted coral reefs, some sponge species are thought to make carbon biologically available by excreting a form of "sponge poop" that other organisms feed on, thereby fueling productivity throughout the ecosystem. In this way, sponges protect the reef against extreme fluctuations in nutrient density, temperature, and light, benefiting the survival of other reef organisms.

A sponge's skeletal type adapts well to its particular habitat, allowing it to live on hard, rocky surfaces or soft sediments such as sand and mud. Some sponges even attach themselves to floating debris! Rarely are they found completely free-floating.

As water filters through a sponge's porous exterior, the sponge gains some motion, receives food and oxygen, and dispels waste. Inside the sponge, tiny hair-like structures called flagella create currents to filter bacteria out of the sponge's cells and trap food within them. Their strong skeletal structures help sponges withstand the high volume of water that flows through them each day.

Marine Cnidarians

Cnidaria is a phylum containing over 9,000 species found only in aquatic and mostly marine environments. All cnidarians have radial symmetrical. There are two major body forms among the Cnidaria - the polyp and the medusa. Sea anemones and corals have the polyp form, while jellyfish are typical medusae. When you look at them, you can see that these body forms are the same except that one is upside down.

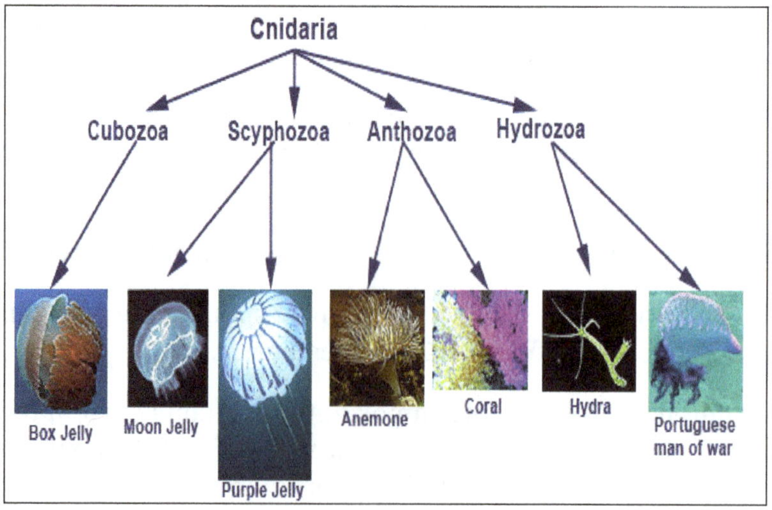

Cnidarians are classified into four main groups: non-moving (sessile) Anthozoa (sea anemones, corals, sea pens); Cubozoa (box jellies); swimming Scyphozoa (jellyfish); and Hydrozoa, a broad group that includes all the freshwater cnidarians as well as many marine forms. There are about 10, 000 species of Cnidarians.

Cnidaria have no organs like hearts or lungs. They have an internal cavity used for respiration and a gastrovascular cavity (a "stomach") with a mouth but they do not have an anus. They possess a body wall made from two layers (the ectoderm and the endoderm) separated by a jelly-like layer called the mesoglea. Other important functions like respiration and excretion occur by diffusion.

Diffusion is the movement of a material (red circles) from regions of higher concentration (Side A) to regions of lower concentration (Side B) until equilibrium is reached (same number of particles move from Side A to Side B as move from Side B to Side A). Diffusion is a natural process caused by the random movement of particles but is only useful to very small (or thin) organisms (ones which have a large surface area compared to their volume) where the material in higher concentration does not have far to move.

Cnidaria do not have a brain or groups of nerve cells ("ganglia"). The nervous system is a decentralized network ('nerve net'), with one or two nets present. They do not have a head, but they have a mouth, surrounded by a crown of tentacles. The tentacles are covered with stinging cells (nematocysts).

Figure: Body form of Cnidaria.

Nearly all Cnidaria are predators, their nematocysts can paralyze and kill prey much larger than them. Small sensory hairs near the nematocysts are sensitive to vibrations in the water. Any prey swimming past can trigger the nematocyst which shoots out the barb. This penetrates the prey's outer covering and injects it with venom. The prey is then moved to the mouth by a tentacle.

Their prey can range in size from plankton to animals several times larger than themselves. Some obtain their energy from algae that that live in their bodies and a few are parasites. Other Cnidaria, including the corals, get their nutrients from symbiotic algae within their cells.

Predators of Cnidaria include sea slugs, sea stars (for example, the Crown of Thorn which can devastate coral reefs), nudibranchs, fish including butterfly fish and parrot fish, which eat corals and marine turtles and sunfish, which eat jellyfish. Plastic bags floating in the oceans are dangerous to turtles which often can mistake them for jellyfish and their guts can become blocked by them.

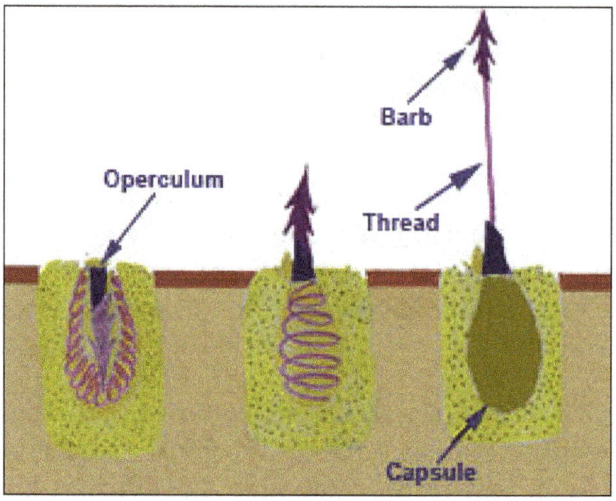

Figure: This is how a nematocyst works.

For example coral, inside the sac of each coral polyp lives single-celled algae called zooxanthellae. The algae produces oxygen and energy (sugars) that the coral polyp needs to live and, in return, the polyp produces carbon dioxide and other substances the algae needs. That is why coral reefs grow so near the surface of the water where the algae can get sunshine for photosynthesis.

Many Cnidaria live in colonies made up of large numbers of individuals joined together in some way. These individuals (called zooids) can either be directly connected by tissues or share a common exoskeleton made from chiton or calcium carbonate.

In the Cnidaria sexual reproduction often involves a complex life cycle with both polyp and medusa stages. For example in Scyphozoa (jellyfish) and Cubozoa (box jellies) a larva swims until it finds a good site, and then becomes a polyp. This grows normally but then absorbs its tentacles and splits horizontally into a series of disks that become juvenile medusae, a process called strobilation. The juveniles swim off and slowly grow to maturity, while the polyp re-grows and may continue strobilating periodically. The adults have gonads in the gastroderm, and these release ova and sperm into the water in the breeding season.

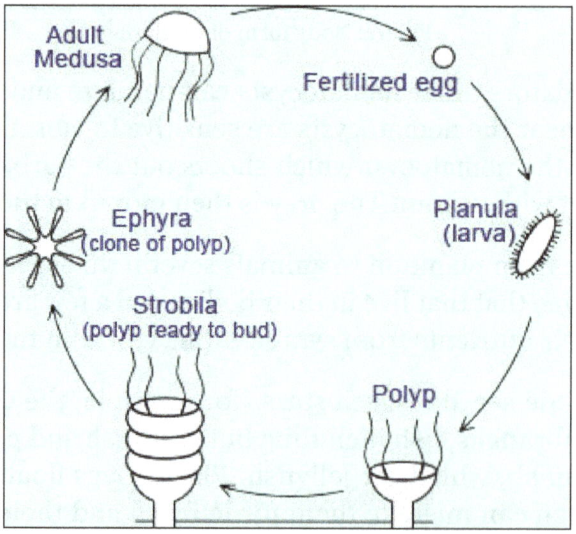

Figure: Life cycle of a jellyfish.

Other Cnidaria have shorter forms of this cycle, for example cubozoan polyps have only one medusa stage. All Cnidaria can reproduce asexually by various means, in addition to regenerating if their bodies are divided into segments or are attacked by predators. Some produce buds, others divide down the middle, others do both. Cnidarians range in size from Hydra, 5–20 millimeters to the Lion's mane jellyfish, which may exceed 2 metres in diameter and 75 metres in length.

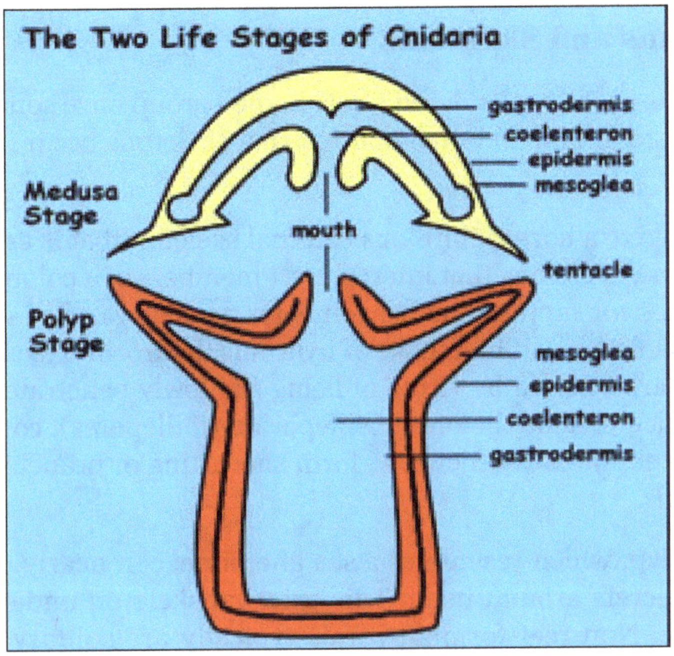

Figure: Structure of Cnidaria.

Tissues and Muscles

Cnidarians consist of two cell layers: an outer ectoderm and an inner endoderm (the gastrodermis) that lines the coelenteron. Between these is sandwiched the mesoglea, a largely noncellular layer composed of a jellylike material permeated by a complex network of supporting fibres that may be microscopically thin or very thick. The fibres and jelly are elastic. In medusae, mesoglea comprises the bulk of the animal and forms a resilient skeleton. In polyps, the water-filled coelenteron acts as a hydrostatic skeleton, which, in concert with the mesoglea, maintains the form of these animals.

Muscles in cnidarians are extensions of the bases of ectodermal and endodermal cells. Individual muscle cells are relatively long and may occur in dense tracts in jellyfish or sea anemones. Most cnidarian muscles, however, are thin sheets at the base of ectodermal and endodermal layers.

In polyps, ectodermal muscles are oriented lengthwise along the cylindrical body and tentacles; endodermal ones are usually circular. Contraction of circular muscles against coelenteric fluid causes the polyp's body to elongate; contraction of longitudinal muscles causes it to shorten. Similar layers of muscles extend and contract the tentacles. Bending results from unequal contractions of longitudinal muscles on opposite sides of the body.

In medusae, all muscles are ectodermal, restricted to the concave oral surface (subumbrellar surface), and organized into circular and radial tracts. Contraction of circular muscles squeezes the subumbrellar space, forcing out contained water and causing the medusa to move by jet propulsion. Recovery of the elastic mesoglea re-extends contracted muscle fibres. Radial muscle contraction distorts the bell, directing the water jet at angles to the longitudinal axis of the bell, allowing the medusa to steer.

Support Mechanisms and Skeletons

Most members of what is usually considered a soft-bodied group have some sort of skeleton aside from the hydrostatic system. Both external and internal skeletons occur in the phylum, but only among polyps.

Most hydroid polyps secrete a horny, chitinous external skeleton that is essentially a tube around the polyp and the network of stolons that interconnect members of a colony. As well as being protective, it confers stiffness for support and has joints for flexibility. A few scyphozoan polyps have comparable chitinous skeletons. Unlike those of hydroids, hydrocoral skeletons are composed of calcium carbonate and are internal by virtue of being shallowly penetrated by channels of living tissue. Hydrocorals, which include the order Milleporina (millepores), commonly called fire coral, and the precious red coral used for jewelry, form encrusting or branching skeletons similar to those of anthozoan corals.

An anthozoan coral polyp, which resembles a sea anemone, can nearly completely retract into the calcareous cup it secrets around itself. This external skeleton underlies a continuous, superficial layer of tissue. Non-reef-forming corals typically are solitary or form small, rather delicately branched colonies, their polyps being relatively large and widely spaced. In some species of reef-forming corals, polyps are so tightly packed that their individual units share common walls. Skeletons may be encrusting, massive, or arborescent (treelike). The latter type of skeleton is delicate and typical of quiet waters at depth or in lagoons, while the former two predominate where water motion is strong. Skeleton is laid down in massive corals at a rate of about one centimetre per year; branching corals may grow considerably more rapidly. The largest corals represent cooperative efforts of up to 1,000,000 tiny individuals precipitating calcium carbonate over centuries. Few attain such proportions, however, and even the largest are eventually broken down by boring organisms such as algae, worms, sponges, and barnacles, as well as by physical processes.

The last major category of cnidarian skeletons, formed by the anthozoan subclass Alcyonaria and the order Antipatharia, are internal. Sea fan and sea whip skeletons consist of the horny protein gorgonin with calcareous spicules fused to form a solid or jointed central rod. Soft coral spicules are discrete, mostly microscopic objects of diverse shapes that vary from needle-like to club- and anchor-shaped. Located in the ectoderm, spicules stiffen the colony. In some species the several spicules that form a protective cup around each polyp may be several millimetres long. The alcyonarian Tubipora is known as the organ-pipe coral after the form of its red calcareous skeleton. Blue corals (the order Helioporacea) have skeletons of crystalline calcareous fibres fused into sheets, which are used for jewelry. Colonies of black coral resemble bushes and may stand more than three metres tall. Their skeletons, made entirely of proteinaceous material similar to gorgonin, are likewise used for jewelry.

Figure: Alcyonaria (soft coral) , Red Sea.

Sea anemones do not produce hard skeletons, although their close relatives in the order Zoanthinaria incorporate foreign objects (sand grains, sponge spicules) into their body walls, which gives them rigidity and toughness. Small anemones that live high in the intertidal zone commonly inhabit abandoned barnacle tests (shells), thereby acquiring some of the benefits of a skeleton.

Nervous System and Organs of Sensation

Medusae have a more highly developed nerve net than do polyps, a feature that is associated with the more active way of life of medusae. Swimming is coordinated by the nervous system. Nervous systems that are capable of conducting nerve impulses both quickly and slowly give these animals considerable behavioral responsiveness and flexibility. Ganglia or other accumulations of nerve cell bodies are not found in cnidarians, but there are gap junctions between neurons and between neurons and effectors, which allow the transmission of nerve impulses. Statocysts, located between the tentacles or near the tentacular base, inform the animal of its orientation with respect to gravitational forces. Light-sensitive ocelli (external patches of pigment and photoreceptor cells organized in either a flat disk or a pit) occur in some medusae of each of the three classes that possess this stage. Such sensory structures are closely associated with a nerve net.

In the past nematocysts were considered independent effectors; that is, they were thought to fire upon appropriate stimulation, without mediation by the nervous system. Evidence, however, favors there being some organismal control over their firing, which may consist only of adjusting the threshold for firing, or the selectivity. Some scientists believe that nervous complexes associated with batteries (groups) of cnidae are the mechanism of control.

Digestion, Respiration and Excretion

Food is taken in and wastes are discharged through the mouth. Extracellular digestion occurs in the coelenteron, which has, in all except hydrozoans and some tiny members of the other classes, radial projections of the wall into the coelenteron that increase the surface area. Ingested material is broken down somewhat in the coelenteron and then taken up by endodermal cells for final intracellular digestion.

Respiration and excretion in cnidarians are carried on by individual cells that obtain their oxygen directly from water—either that in the coelenteron or that of the environment—and return metabolic wastes to it. Thus, all physiological functions are carried out at no more than the tissue level of differentiation.

Defense and Aggression: Nematocysts

Cnidae range from only about 10 to 100 micrometres (0.0004 to 0.004 inch) long, but they are among the most complex intracellular secretion products known. Each consists of a spherical or cigar-shaped capsule with an eversible, hollow tubule extending from one end. In the unfired state, the tubule is coiled within the capsule. When a cnidarian contacts a predator or prey item, the capsule opens and the tubule everts. The tubule may be adhesive, or it may entangle the object. Both types serve to hold food items. A third type of tubule is armed with spines that penetrate predator or prey. Toxins contained in the capsule are injected through the tubule into the object being held. Each Cnida can be fired only once. Undifferentiated interstitial cells of the ectoderm and endoderm appear to be the source of the cnidoblasts (cells that produce cnidae).

Evolution

The exact relationships between the different cnidarian groups are unknown. Among theories proposed on the evolution of the phylum Cnidaria, most treat the radial symmetry and tissue level of organization as evidence that the group is primitive (that is, it evolved before the evolution of bilateral symmetry) and hold that the medusa is the original body form, being the sexually reproductive phase of the life cycle. Another theory is that the original cnidarian was a planula-like organism that preceded both polyp and medusa. In either case, Hydrozoa is considered to be the most ancient of cnidarian classes, and Trachylina is thought to be the most primitive extant order of that group. An alternative view is that anthozoans are the stem of the phylum, which evolved from bilateral flatworms and is secondarily simplified. A corollary to this theory is that the polyp is the ancestral body form.

Speculations about the origin of the phylum are not easily resolved, for preservable skeletal structures developed relatively late in cnidarian evolution. The oldest fossilized cnidarians were soft-bodied. Representatives of all four modern classes have been identified in Ediacaran fauna of the Precambrian Period (that is, those appearing between about 635 million and 541 million years ago) known from more than 20 sites worldwide. As much as 70 percent of Ediacaran species have been considered to be cnidarians. Curiously, there are few fossil cnidarians of the Cambrian Period (541 million to 485.4 million years ago). The Conulariida, which existed from the Cambrian Period to the Triassic Period (251.9 million to 201.3 million years ago), are considered by some scientists to be skeletal remains of scyphopolyps, either ancestral to the coronates or without modern derivatives. Presumed fossil sea anemones are found in the lower Cambrian System. Colonies of Stromatoporoidea, considered to be an order of the class Hydrozoa that extended from the mid-Cambrian Period to the Cretaceous Period (about 145 million to 66 million years ago), produced massive skeletons. Although there were two groups of Paleozoic corals, neither of which has modern descendants, they were not great reef-builders during that era. Scleractinians arose in the mid-Triassic Period; blue corals, gorgonians, millepores, and hydrocorals have records from the Jurassic Period (201.3 million to 145.0 million years ago) or the Cretaceous Period to the present. Most other cnidarians are known only from the Holocene Epoch (within the last 11,700 years).

Classification

Annotated Classification

The following classification, limited to living cnidarians, generally follows that used by D.G. Fautin in S.P. Parker (ed.), Synopsis and Classification of Living Organisms and L.H. Hyman, The Invertebrates, Protozoa Through Ctenophora.

Phylum Cnidaria (Coelenterata)

- Nematocyst-bearing, radial metazoans without organs.
- Have a cellular inner endoderm and outer ectoderm, separated by noncellular mesoglea.
- Polyp and medusa forms; either or both may be present in one life history.
- Most polyps have tentacles around mouth; tentacles of medusae at bell margin.
- One internal cavity, the coelenteron, has 1 opening to exterior, the mouth.
- About 9,000 species.

Class Anthozoa

- Exclusively polypoid with biradial symmetry.
- Oral end a disk with central mouth and hollow tentacles arising at margin and/or on surface.
- Mouth leads to coelenteron via stomodaeum that has ciliated troughs (siphonoglyphs) for water transport into and out of coelenteron.
- Coelenteron divided by radial mesenteries that extend inward and insert on the stomodaeum (complete mesenteries) or not (incomplete mesenteries).
- About 6,000 species.

Subclass Alcyonaria

- Octocorals.
- Polyps with 8 pinnately branched tentacles, 8 mesenteries, and a single siphonoglyph.
- Nearly all colonial with internal skeletons.

Order Stolonifera

- Polyps of colony connected by stolons.
- Skeletons of spicules or horny external cuticle.
- Shallow tropical and temperate seas.

Order Telestacea

Long axial polyps bear lateral polyps. Skeleton of spicules fused with a horny material.

Order Gorgonacea

- Sea fans and sea whips.
- Colonies commonly arborescent with axial skeleton of gorgonin and/or calcareous spicules.
- Polyps rarely dimorphic.
- Tropical and subtropical.

Order Alcyonacea

- Soft corals.
- Small to massive colonial forms.
- Lower parts of polyps fused into a fleshy mass; oral ends protrude.
- Internal skeleton of isolated calcareous spicules.
- Primarily tropical.

Order Helioporacea (Coenothecalia)

- Blue coral. Massive lobed calcareous skeleton.
- Tropical Caribbean and Indo-West Pacific species.

Order Pennatulacea

- Sea pens and sea pansies.
- Fleshy, always dimorphic, unbranched colonies, with 1 axial polyp and many lateral ones.
- Polyp-free peduncle burrows into soft sediments; polyp-bearing distal end of the polyp (rachis) extends into water and may be completely retractile.
- Central skeleton a calcified axial rod; polyps and rachis have isolated calcareous spicules.

Subclass Ceriantipatharia

- Black corals and tube anemones.

Order Antipatharia

- Black coral.
- Large bushy colonies with thorny, hornlike axial skeleton formed by small polyps with 6 simple tentacles and 1 siphonoglyph.
- Mostly tropical and subtropical.

Order Ceriantharia

- Tube anemones.
- Solitary polyps with 2 sets of tentacles (oral and marginal) that form feltlike tubes of specialized cnidae (ptychocysts) and burrow in soft sediments.
- Shallow waters worldwide.

Subclass Zoantharia

- Sea anemones and corals.
- Six (or multiples of 6) tentacles (rarely branched).
- Mesenteries commonly arranged hexamerously.
- Solitary or colonial.
- Skeletons non-spicular calcareous, horny, or lacking.
- Usually 2 siphonoglyphs.

Order Actiniaria

- Sea anemones.
- Solitary or clonal, never colonial; lacking skeleton; with or without basilar muscles.
- Mostly littoral or benthic, commonly attached to firm substrata but some burrow in soft sediments, found Worldwide.

Order Corallimorpharia

- Sea-anemone-like solitary or aggregated polyps lacking basilar muscles and skeleton.
- Coral-like muscles and nematocysts.
- Mostly tropical.

Order Ptychodactiaria

- Sea-anemone-like, lacking ciliated tract on edge of mesenteries and basilar muscles.
- Both poles.

Order Scleractinia (Madreporaria)

- True or stony corals.
- Mostly colonial; calcareous external skeleton; no basilar muscles or siphonoglyphs.
- Mostly tropical and subtropical.

Order Zoanthinaria (Zoanthidea)

- Solitary, clonal, or colonial polyps resembling sea anemones.
- Lack skeleton but may incorporate debris into body wall, commonly epizoic.
- One complete and 1 incomplete mesentery per pair, mostly tropical.

Class Cubozoa

- Tropical, cuboidal medusae that swim strongly; box jellyfishes.
- Margin simple with single or grouped tentacles arising above the 4 corners.
- Polypoid stage of most species unknown.
- Fiercely stinging members can cause human fatalities.
- Contains 2 orders, Carybdeida and Chirodropida.

Class Hydrozoa

- Life histories may involve both polypoid and medusoid stages, but either may be suppressed or absent.
- Tetramerous or radially symmetrical medusae small, with shelf of tissue (velum) across lower part of bell, which reduces diameter of subumbrellar aperture (condition known as craspedote).
- Colonial forms commonly polymorphic. Coelenteron undivided. Gametes ripen in ectoderm. Only class with some freshwater members. 2,700 species.

Order Actinulida

- Curious groups of solitary, motile cnidarians with features of both polyps and medusae.
- Europe; in marine sand.

Order Chondrophora

- Floating polymorphic colonies supported by chitinous skeleton.
- Free medusae are produced; includes Velella.
- Oceanic, worldwide.

Order Hydroida

- Hydroids.
- Usually colonial and polymorphic, release free medusae or retain modified medusoid reproductive structures on polyp colony.

- Polyps usually have a chitinous exoskeleton.
- Includes naked, solitary freshwater polyp Hydra.
- Largest order of Hydrozoa.

Suborder Anthomedusae

- Medusae bell-shaped, with gonads on the stomach or sides of manubrium.
- Sensory structures consist of pigmented eyespots (ocelli).
- Skeleton, if present, lacks cup (hydrotheca) into which polyp may withdraw (a condition known as gymnoblastic), few species with calcareous exoskeleton.
- Most abundant in bays and shallow coastal waters.

Suborder Leptomedusae

- Medusae saucer-shaped, but lacking in many species.
- Gonads on radial canals.
- Sensory structures usually statocysts.
- Hydroids with hydrothecae (condition known as calyptoblastic).
- All shallow marine waters.

Suborder Limnomedusae

- Small medusae with gonads on stomach walls or radial canals.
- Polyps solitary or colonial, commonly with 1 or 2 tentacles, and no skeleton.
- Mostly freshwater.

Order Milleporina

- Fire coral.
- Colonial forms producing massive calcareous skeletons.
- Gastrozooids and dactylozooids project through pores in surface of skeleton.
- Reduced, acraspedote (lacking a velum) nonfeeding medusae are released.
- Tropical.

Order Siphonophora

- Pelagic polypoid colonies with greatest degree of polymorphism in phylum, lack medusae.

- Oceanic, worldwide.
- Includes Portuguese man-of-war, Physalia.

Order Stylasterina

- Hydrocorals.
- Resembling millepores, colonies erect and branching or prostrate.
- Commonly yellow, red, or purple.
- Reduced medusae not freed, develop and produce gametes in cavities of skeleton (ampullae).
- Worldwide; includes precious red coral, Corallium.

Order Trachylina

- Medusa dominant, reduced or no polyp stage.
- Statocysts and special sensory structures (tentaculocysts).
- Differ from other hydromedusae by having tentacles inserted above umbrellar margin.
- Oceanic, mostly warmer waters.

Suborder Laingiomedusae

- Medusae with features of both Narcomedusae and Trachymedusae.
- Polyp unknown.

Suborder Narcomedusae

- Scalloped margin, gonads on stomach walls.
- Manubrium lacking.

Suborder Trachymedusae

- Smooth bell margin, gonads on radial canals arising from the stomach.
- Polyp and asexual reproduction absent.

Class Scyphozoa

- Exclusively marine group in which acraspedote medusae predominate.
- Life histories commonly involve alternation of a very small polyp, the scyphistoma, with a medusa, which develops from an ephyra released by the polyp.

- Coelenteron of both divided by 4 longitudinal septa producing tetramerous radial symmetry.
- Gonads endodermal.
- Marginal sensory structures (rhopalia) with statocysts and/or ocelli.
- Most abundant in coastal waters, but oceanic species exist.
- About 200 species.

Order Coronatae

- Large medusae that are conical, dome-shaped, or flattened, with furrow around bell above scalloped margin.
- Some species have scyphistoma stage with external chitinous skeleton.
- Oceanic, some species living at great depths.

Order Rhizostomae

- Medusae like those of Semaeostomeae but with mouth subdivided into minute pores that connect with coelenteron.
- Mostly tropical.
- Deep-water species may lack polypoid stage.

Order Semaeostomeae

- Most common and best known jellyfishes.
- Full alternation of polyp and medusa stages.
- Bell domed or flattened, with the margin scalloped into 8 or more sections.
- Edges of single mouth drawn out into 4 long arms.
- Most species in warm, coastal waters, a few in frigid waters; some oceanic.
- Includes the giant Cyanea arctica, which may attain 2 m in diameter.

Order Stauromedusae

- Sessile jellyfish that are vase-, goblet-, or trumpet-shaped, and usually bear 8 groups of tentacles.
- No more than 2–3 cm long.
- Apparently lacking polypoid stage.
- Temperate and cold temperate waters worldwide.

Marine Worms

Marine worms are worms that live in a marine environment. Marine worms belong to many different phyla, including the Platyhelminthes (flatworms) , Nematoda (roundworms), Annelida (segmented worms including tube worms and bristle worms), Chaetognatha (arrow worms), Hemichordata (acorn worms), Phoronida (horseshoe worms), Nemertea (ribbon worms or proboscis worms), Sipunculid worms (peanut worms) and Echiura (Spoon Worms and Innkeeper Worms). Marine worms are found living under boulders of rocky ocean shores, among the holdfasts of algae, in mud, sandy mud and in sand, while others build tubes in which to shelter. They can be carnviores, herbivores or parasites. They are found at all depths including hydrothermal vents and hydrocarbon seeps in the depths of the oceans. Marine worms may only have their shape and the fact that they have 'worm' in their name in common.

Role Marine Worms Play in the Ecosystem

A saltwater invertebrate spanning more than 10 separate phyla, the marine worm, plays a vital role in the development and sustainability of many of the planet's watery ecosystems. Adding to the thousands of known species, the scientific community continues to discover new worms and worm behavior, not previously recorded.

Worm Role

The relationship between the marine worm and its habitats is symbiotic. The coral reef thrives from the worm's eating, building, reproduction and excretion habits, while the worm sustains itself on reef particulates and abides the security provided from the reef's rocky ridges. Not all marine worms live in the oceans, but reside in shoreline tide pools, or the brackish water out-crops of the lowlands. Here the marine worms bury themselves in the sediment. The tunneling and boring of marine worms irrigate and oxygenate the shallow water pools encouraging beneficial plant and algae growth. Whether it's in tide pools, lowland waters or oceanic reefs, the marine worm's primary ecological contribution is as sustenance for aquatic animals further up the food chain.

Reproduction

Marine worms have distinct sexes and ephemeral sex organs. Tiers of gonads amass inside of marine worms to produce either sperm or eggs. When the fertilizer or eggs are ready to be released, gonad ducts are formed but are eliminated by absorption after reproduction. They emerge in a larval stage, shaped like a saucer, and join the rest of the free-floating sea plankton. As plankton, the marine worms provide nutritional sustenance to a great number of the ocean's inhabitants. If the larva survives the open ocean, a miniature worm develops inside a floating sac-like encasing. Once fully formed, the marine worm eats its way out of the sac and falls to the ocean floor, where it begins the eco-cycle all over again. More egg sacs are devoured than survive, keeping marine worms to a manageable number.

Nutritional Needs

The marine worm resides at the bottom of the oceanic food chain, and many animals in the ocean feed on them. The marine worm feeds primarily on oceanic particulate. However, marine worms are accomplished predators and eat fish, snails, crustaceans and clams. They can catch snails or bivalves and wrap their bodies around their victims and inject them with digestive enzymes. The marine worm then excretes nutritional excrement that nourishes and acts as compost, for surrounding microorganisms, bacterium and plant life.

Ecological Identity

Most marine worms fall under one of four habitually descriptive classes including tube-dwelling, boring, commensalism and parasitism. Like an oceanic barometer, the study of marine worms often gives scientists advanced notice to potential problems in aquatic habitats. Biological monitoring and mapping of marine worms can determine if hazardous aquatic pollutants, or biologically dangerous bacteria, can be detected. Other biodiversity is uncloaked at nominal levels from the study of sea worms, such as disease-causing microorganisms or physical disturbances radical enough to upset the natural function of the planet's marine reserves.

Marine Worms - Annelids (Segmented worms)

The annelids or Annelida (also called "ringed worms"), are a large phylum of segmented worms, with over 17,000 modern species including polychaetes, clitellates, ragworms, earthworms and leeches. They are found in most marine environments from tidal zones to hydrothermal vents, in freshwater, and in moist terrestrial environments.

The basic annelid body consists of multiple segments, each of which has the same sets of organs. They are soft-bodied invertebrates with long, narrow bodies. They possess a digestive system, a circulatory system and a nervous system. Their bodies are covered by an outer covering (cuticle) made of tough but flexible collagen.

Polychaetes (Tube Worms)

Polychaetes (meaning many bristles) are (often) brightly coloured segmented worms. Most are less than 10 centimeters long, although they can range from 1 millimeter to 3 metres. There are about 12,000 species of Polychaetes.

Each segment has a pair of paddle-like fleshy paired appendages (parapodia), used for movement and, in many species, for respiration. Bundles of bristles (setae) are found on the parapodia. Their body structures can vary according to their lifestyles which include crawling along the bottom, burrowing, free swimming, tube-dwelling or boring, commensalism and parasitism.

The well-developed head usually has two to four pair of simple eyes capable of distinguishing only light and dark. It also has a pair of antennae, pair of feelers (palps) and a pair of pits lined with cilia which act as chemoreceptors. These help the polychaetes find food.

Their mouths vary depending on their diet, since the group includes predators, herbivores, filter feeders, scavengers, and parasites. Most have a pair of jaws and a pharynx that can be quickly turned inside out, allowing the worm to grab food and pull it into the mouth.

Figure: Tube worm.

Figure: Tube worm.

Most species have external gills usually found on the parapodia and a well-developed, simple circulatory system. The nervous system consists of a single or double ventral nerve cord running the length of the body, with ganglia and a series of small nerves in each segment. The brain is relatively large, compared with that of other annelids, and lies in the upper part of the head.

Most polychaetes have separate sexes - male and female and the sperm and eggs are released into the surrounding water through ducts or openings. The fertilised eggs hatch into larvae, which float among the plankton, and eventually metamorphose into the adult form by adding segments.

Echinoderms

Many echinoderms actually do have "spiny" skin, but others do not. This phylum exists exclusively in the sea, and cannot be found on land or in fresh water. All echinoderms have one thing in common: radial symmetry. This means that the creatures have appendages (or body construction) which point outward from the center of the body like the spokes on a bicycle wheel. Furthermore, these appendages usually occur in multiples of five, although there are a few exceptions. There are several well-known members of this group, like sea stars and sea urchins. The radial symmetry is obvious in these creatures.

Perhaps not as obvious is the water vascular system, another trait common to all echinoderms. By examining the oral (ventral or underside) side of a sea star, one will be able to see hundreds of tiny feet usually arranged into several rows on each ray (appendage) of the star. These are called tube feet, or podia, and are filled with sea water in most echinoderms. The water vascular system within the body of the animal is also filled with sea water. By expanding and contracting chambers within the water vascular system, the echinoderm can force water into certain tube feet to extend them. The animal has muscles in the tube feet which are used to retract them. By expanding and retracting the right tube feet in the proper order, the creature can walk. Many echinoderms can also form suckers on the ends of their tube feet. These suckers can be used to capture and hold prey, or to hold onto rocks in a swift current or tide.

Interestingly, although most mature echinoderms are benthic (meaning that they live on the bottom), the larvae are usually planktonic with bilateral symmetry. During the process of maturing, the echinoderm will change its body shape from bilaterally symmetrical to radially symmetrical, and in the process, settle down on the sea floor.

Figure: The Cushion Star, found in the Caribbean, is a classic example of an Asteroid.

The Stelleroids

Perhaps the most common echinoderm is the sea star. Although more well known as the star fish, sea stars are not actually fish. The scientific community prefers to reserve the term "fish" for vertebrates with fins. The subphylum Stelleroidea contains the two classes of sea stars. The class Asteroidea contains the true sea stars and sun stars. The class Ophiuroidea contains the brittle stars and basket stars. The distinction between the two classes lies in the way the arms connect with the body. Ophiuroids have a distinct central body part (called a central disk) with arms radiating out from the body. Adjacent arms do not connect with each other. Asteroids, on the other hand, have arms which seem to connect together in such a way as to make it difficult to discern where the arms end and the central disk begins.

Figure: This Northern sea star, photographed in Massachusetts, is regenerating two lost arms. It will take over a year, but they will be good as new.

The sea star's aboral (top) surface is spiny looking when closely examined. The rumpled skin, when examined up close, contains several different types of formations. Some of the bumps on the surface are called Dermal Branchiae, which are used to absorb oxygen from the water. Another type of adaptation of the skin are called pedicellaria. These pincher-like pairs of organs can be used to pluck things off of the skin of the sea star which might otherwise cause a problem. For example, the larval form of a barnacle could settle onto a sea star and begin to grow if the sea star had no way of removing it.

Figure: The voracious Spiny Sun star of the North Atlantic stalks and eats other sea stars! It's an Asteroid with up to 12 arms.

The single colored spot on the aboral surface of the sea star is called the madreporite. This is a calcareous piece of the water vascular system filled with tiny holes, like a strainer. It is the interface between the water vascular system and the ocean, acting as a filter.

The sea star has a light sensitive organ at the tip of each ray, called an eyespot. When moving across the ocean floor, the sea star usually leads with one ray, probing the surface ahead. Although the star cannot see in the way we do, it can detect the presence and direction of light, and does seem to have some idea where it is headed.

Sea stars are capable of regenerating limbs in the event that one or more is severed or damaged. The wound first closes off, and in time, the new limb will begin to grow. In a few species, the severed limb can regenerate a new sea star, but in most species, the severed limb dies.Sea stars eat a variety of different things, including barnacles, clams, mussels, snails, sea urchins, and in some cases, other sea stars! Many sea stars, such as the Northern Sea Star, eat mussels and clams in a fascinating way. The sea star first surrounds its intended victim. Then it applies outward force (with its suction cup equipped tube feet) on the two mussel shells (called valves), to pull them apart. Contrary to popular belief, the sea star does not need to apply force for a long time in order to tire out the mussel. The sea star can apply so much force to the mussel valves (7 or more pounds!) that it will bend the shell. Seizing the moment, the sea star then everts its stomach out through its mouth, and into the mussel (only a 1/100th of an inch opening is required). Once the sea star begins to digest its victim within the victim's own body, the victim dies. The sea star then finishes the meal by consuming the rest of the mussel. When the star is finished with the mussel, nothing remains but a shell.

Echinoids

The class Echinoidea includes the sea urchins, heart urchins, cake urchins and familiar sand dollars. The sea urchins are well known to most people, having a large number of sharp spines pointing out in all directions. These spines offer protection from many would-be predators. The spines are joined to the skeleton of the animal, called the "test," in a form of ball-joint. This coupled with the fact that there are muscles attached to each spine enables the urchin to swivel its spines in the direction of a predator. The test is an egg-like spherical structure constructed of rows of radially arranged plates fused together.

Figure: This Purple sea urchin from California has hundreds of pointy spines for protection.

The creature has 5 paired rows of tube feet which, when extended, are long enough to reach past the length of the spines. They contain suckers. The anus is on the top of the creature, while the mouth is on the underside. The mouth contains five teeth which are arranged pointing towards the center of the mouth. This structure looks and works like the jaws of a drill chuck and has been termed "Aristotle's Lantern" because it was first described by Aristotle as looking like the top of an oil lamp. The creature uses its tube feet to pull itself against the substrate so it can gnaw away at algae with its mouth.

Holothuroids

The class Holothuroidea is composed of creatures called sea cucumbers. A sea cucumber is so named due to the fact that many members of the group (but not all) resemble the garden variety of cucumber. The similarity ends there.

Figure: The Orange-footed sea cucumber lives in the North Atlantic. It resembles a football with tentacles at one end for feeding.

Sea cucumbers are usually somewhat football shaped and lay on their side on the bottom. They have 5 rows of tube feet running lengthwise, like the seams on a football. Three of the rows of tube feet are well developed and are in contact with the substrate. The other two rows are usually either underdeveloped (and not used) or missing completely. At the oral end of the animal is a mouth surrounded by tentacles. These tentacles, usually branched, are actually special tube feet and are part of the water vascular system. The water vascular system is not filled with sea water in sea cucumbers, as in sea stars and urchins, but rather with a special body fluid. There is, therefore, no direct interface (madreporite) between the outside water and the internal organs of the animal.

Figure: The Tiger Tail sea cucumber feeds on particles found in the sand on the bottom.

The sea cucumber feeds in a fascinating way. It may position itself in a spot on the ocean floor where a current will bring a steady supply of food (plankton and other organic particles) its way. The tentacles are opened to collect the food. Then, the cucumber sticks each tentacle in its mouth, one at a time, and licks them off. As soon as a tentacle has been licked, the cucumber takes it out of its mouth and holds it out to collect more food while it licks the next tentacle. The sea cucumber will do this for hours at a time. Other kinds of sea cucumbers use their tentacles to sift through sand on the bottom for particles of food, rather than collect plankton from the water.

Many sea cucumbers are quite poisonous. These cucumbers, if injured, can kill fish in the same aquarium. The poison of some sea cucumbers has shown promise as an inhibitor in the growth of cancer cells.

The sea cucumber has another interesting (if not slightly disgusting) habit, called self-evisceration. When a sea cucumber is attacked, it may expel some of its internal organs. This could either satisfy a predator, or scare it off. The cucumber will then proceed to grow another set of organs.

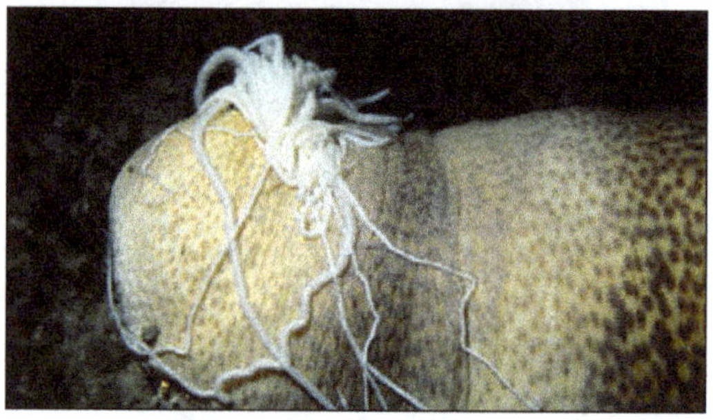

Figure: This sea cucumber is eviscerating a portion of its intestines to thwart an attack.

Some sea cucumbers can secrete a sticky glue-like substance as a defense mechanism. This goo is so sticky that it absolutely cannot be removed from the skin without shaving off any hair with which it has come in contact. Historically, people have used this substance as a bandage to bind wounds.

The Chinese eat certain sea cucumbers, and consider them quite delicious. Some species have edible muscular body walls. No matter what you think about the looks of Echinoderms, they are very interesting and diverse animals with amazing adaptability.

Marine Molluscs

Molluscs are the second species-rich phylum in the world after the arthropods. This is also true in subterranean ecosystems, where they are sometimes common in aquatic habitats, and more occasionally in terrestrial habitats. Nearly all obligate cave-dwelling molluscs are gastropods. Morphological characteristics of obligate subterranean species include a thin, often translucent shell that is usually white, with a depigmented body, and depigmented and reduced eyes. The aquatic species of gastropods are usually very tiny, and a lot of species are currently being described. There appear to be two hotspots of biodiversity—the Mediterranean region in Europe, mostly for stygobite species, and East Asia, where most of the troglobite species live. Subterranean molluscs are deposit feeders and the species richness in caves depends on the input of organic matter.

Mollusc have extreme variation in reproductive characteristics. A generic mollusc is dioecious with paired gonads. Eggs or sperm are released into the coelomic cavity and carried outside by ducts. Fertilization is external and the zygote develops into a gastrula and then a free-swimming trochophore larva. Trochophore larvae develop into a veliger in marine taxa. The veliger larvae have a well-developed foot, shell, and other structures. A veliger will settle on the bottom of the water body and metamorphose into the adult form. In freshwater unionid bivalves, larvae develop into glochidia that are brooded and later released to attach to fish hosts used for dispersal; others have direct development into adults (Sphaeriidae or pill clams); while some retain the veliger larval form (e.g., Dreissena or zebra mussels). Freshwater gastropods also have great variation in developmental patterns. Gastropods can be oviparous, ovoviviparous, or viviparous. Larval forms are present in some taxa. For example, the neritids are oviparous and deposit eggs on substrates that hatch into a swimming larval form. The larvae undergo metamorphosis into the adult form. Oviparous taxa have direct development, while ovoviviparous species brood their young and release them at a larger size.

The molluscs (also called mollusks) compose the large phylum of invertebrate animals known as the Mollusca. Around 85,000 extant species of molluscs are recognized. Molluscs are the largest marine phylum, comprising about 23% of all the named marine organisms. Numerous molluscs also live in freshwater and terrestrial habitats. They are highly diverse, not just in size and in anatomical structure, but also in behavior and in habitat.

Marine molluscs (eg, Opisthobranchia, E. rufescens) are filter feeders (ie, they feed by filtering seawater). Because seawater contains high concentrations of bacteria, these organisms produce within themselves powerful chemicals to defend themselves from potentially harmful microorganisms.

The chemicals produced by the molluscs are primarily Kahalalides, which are bioactive peptides and anticancer agents. Garcia-Rocha et al. found that the antitumoral compound Kahalalide-F acts on cell lysosomes. Note that lysosome is a membrane-bound cell organelle found in most animal cells, but are absent in red blood cells. While Suarez et al. found that Kahalalide-F induces oncosis (a condition characterized by the formation of one or more tumors) in human prostate and breast cancer cells, Janmaat et al. discovered that this marine-derived compound induces necrosis-like cell death that involves depletion of ErbB3 and inhibition of Akt signaling. While Goetz et al. and Bonnard et al. determined the stereochemistry of Kahalalide-F, Lopez-Macia et al. carried out its synthesis and structure determination. Sewell et al. identified the mechanism of action of Kahalalide-F, particularly variable cell permeability in human hepatoma cell lines.

Medical application of any drug in patients demands its preclinical and clinical studies. Preclinical toxicity studies of Kahalalide-F as a new anticancer agent were carried out by Brown et al. While Ciruelos et al. carried out phase-I clinical and pharmacokinetic study with Kahalalide-F in patients with advanced solid tumors with a continuous weekly 1-h iv infusion schedule, Rademaker-Lakhai et al. carried out its phase-I clinical and pharmacokinetic study in patients with advanced androgen refractory prostate cancer.

Marine Arthropods

Arthropods are organisms belonging to the kingdom Animalia and the phylum Arthropoda. They are a very diverse group of animals that includes but is far from limited to insects, crustaceans, spiders, scorpions, and centipedes. Arthropods make up the largest phylum in the world, with more numbers and species diversity than most other phyla. With more than 800,000 known species of arthropods, it is no wonder that they dominate the land and sea.

Characteristics of Arthropods

All Arthropods

- Jointed legs: Jointed legs allow arthropods to travel quickly regardless of their method of transportation. Whether swimming or scurrying across the ground, arthropods are speedy because of their jointed legs.
- A segmented body: An arthropod's body can be divided into one, two, or three main sections. If they have one section, it is called a trunk. If they have two sections, these are called the cephalothorax and abdomen. If they have three sections, the third section is the head.
- A hard exoskeleton: The exoskeleton of an arthropod is made of a strong polysaccharide called chitin. This hard shell protects the animal, retains moisture, and sometimes even plays a role in reproduction.
- Compound eyes: Compound eyes allow arthropods to take in their environment in a variety of ways. Arthropods can see through a very wide lens and use their compound eyes to detect the slightest of motions and perceive any depth.

Additional features make certain species of arthropods better suited for their specific habitat.

Terrestrial Arthropods

Land dwelling arthropods have a number of features that enable them to succeed in their environment.

- Stinger: The stinger allows terrestrial arthropods to inject their prey with poison and paralyze, injure, or dissolve it into an edible liquid.
- Book Lungs/Trachea: In order to respirate air, terrestrial arthropods need a special set of lungs and/or trachea. Book lungs are layered organs that expand to take air in and contract to absorb it.
- Spinnerets: Terrestrial arthropods like spiders use spinnerets to produce webs. These can be used for shelter, prey entrapment, courtship, etc.

Aquatic Arthropods

Like land-dwelling arthropods, aquatic arthropods require adaptations that make living wholly or partially underwater possible.

- Gills: Just as book lungs allow for terrestrial respiration, gills allow for aquatic respiration. Marine arthropods use their gills to take in water and absorb its oxygen into their bloodstream.
- Cement Glands: Cement glands are unique adaptations that allow barnacles to adhere to nearly any surface. The adhesive secreted helps barnacles cling to rocks, ships, and other organisms and is so strong that scientists study its properties as inspiration for new materials.
- Swimmerets: Swimmerets allow some species of aquatic arthropods to swim, a movement that closely resembles running quickly through the water. In some species, a pair of swimmerets is used to inseminate mates.

Habitat and Distribution

Arthropods can survive in nearly any habitat. Different species can be found on dry land, water, or a combination of both. Aquatic arthropods are often found in coastal habitats like sandy beaches and intertidal areas but can even live comfortably in the deep sea. Horseshoe crabs are one of the oldest known species of marine arthropods. They have been known to inhabit both deep ocean waters and coastal sands. With as many species of arthropods as are living on the earth, it is far more difficult to find an environment or ecosystem where arthropods are not present than to find one where they are.

Reproduction

Arthropods usually reproduce sexually through external fertilization or, more uncommonly, asexually in cases where both male and female reproductive organs are present in one organism. External fertilization occurs when a male arthropod encases its sperm in a pouch that is deposited directly into a female arthropod or sent free to be taken up by a female.

The offspring of most species of arthropods begin as eggs, then hatch from these and enter a larval stage. In many arthropods, such as crabs, you can see these eggs attached to the hard abdomen. The larvae undergo metamorphosis, sometimes emerging from a cocoon during the pupal stage, to progress into adulthood. Water presents interesting challenges to the offspring of aquatic arthropods. Throughout this process of metamorphosis, young marine arthropods drift through the sea and can cover great distances in this manner. They have no control over where they end up before they reach adulthood.

Examples of Marine Arthropods

Examples of marine arthropods include:

- Lobsters.
- Horseshoe crabs.
- Barnacles.
- Isopods.
- Skeleton shrimp.
- Krill.
- Crabs (e.g., green crab, spider crab, hermit crab).
- Sea spiders.
- Copepod.
- Amphipods.
- Barnacles.

References

- Evolution, cnidarian, animal: britannica.com, Retrieved 25, June 2020
- Role-marine-worms-play-ecosystem-11067: animals.mom.com, Retrieved 13, August 2020
- Marine-worms02, marine-worms: mesa.edu.au, Retrieved 01, March 2020
- Echinoderm, wonders, education: oceanicresearch.org, Retrieved 08, July 2020
- Mollusc, earth-and-planetary-sciences: sciencedirect.com, Retrieved 12, January 2020
- Marine-arthropod-facts-2291818: thoughtco.com, Retrieved 06, April 2020

Chapter 4

A Comprehensive Study of Marine Habitats

Marine habitats are the ecological or environmental areas that support marine life. Marine habitats form an important part of marine ecosystems as they are rich in biodiversity and support a wide variety of life forms. Marine habitats can be further divided into coastal and open sea habitats. This chapter delves into the realm of marine habitats and provides an in-depth explanation to readers.

The marine environment supplies many kinds of habitats that support marine life. Marine life depends in some way on the saltwater that is in the sea (the term *marine* comes from the Latin *mare*, meaning sea or ocean). A habitat is an ecological or environmental area inhabited by one or more living species.

Marine habitats can be divided into coastal and open ocean habitats. Coastal habitats are found in the area that extends from as far as the tide comes in on the shoreline out to the edge of the continental shelf. Most marine life is found in coastal habitats, even though the shelf area occupies only seven percent of the total ocean area. Open ocean habitats are found in the deep ocean beyond the edge of the continental shelf.

Alternatively, marine habitats can be divided into pelagic and demersal habitats. Pelagic habitats are found near the surface or in the open water column, away from the bottom of the ocean. Demersal habitats are near or on the bottom of the ocean. An organism living in a pelagic habitat is said to be a pelagic organism, as in pelagic fish. Similarly, an organism living in a demersal habitat is said to be a demersal organism, as in demersal fish. Pelagic habitats are intrinsically shifting and ephemeral, depending on what ocean currents are doing.

Marine habitats can be modified by their inhabitants. Some marine organisms, like corals, kelp, mangroves and seagrasses, are ecosystem engineers which reshape the marine environment to the point where they create further habitat for other organisms.

Only 29 percent of the world surface is land. The rest is ocean, home to the marine habitats. The oceans are nearly four kilometres deep on average and are fringed with coastlines that run for nearly 380,000 kilometres.

In contrast to terrestrial habitats, marine habitats are shifting and ephemeral. Swimming organisms find areas by the edge of a continental shelf a good habitat, but only while upwellings bring nutrient rich water to the surface. Shellfish find habitat on sandy beaches, but storms, tides and currents mean their habitat continually reinvents itself.

The presence of seawater is common to all marine habitats. Beyond that many other things determine whether a marine area makes a good habitat and the type of habitat it makes. For example:

- Temperature – is affected by geographical latitude, ocean currents, weather, the discharge of rivers, and by the presence of hydrothermal vents or cold seeps.
- Sunlight – photosynthetic processes depend on how deep and turbid the water is.
- Nutrients – are transported by ocean currents to different marine habitats from land run-off, or by upwellings from the deep sea, or they sink though the sea as marine snow.
- Salinity – varies, particularly in estuaries or near river deltas, or by hydrothermal vents.
- Dissolved gases – oxygen levels in particular, can be increased by wave actions and decreased during algal blooms.
- Acidity – this is partly to do with dissolved gases above, since the acidity of the ocean is largely controlled by how much carbon dioxide is in the water.
- Turbulence – ocean waves, fast currents and the agitation of water affect the nature of habitats.
- Cover – the availability of cover such as the adjacency of the sea bottom, or the presence of floating objects.
- The occupying organisms themselves – since organisms modify their habitats by the act of occupying them, and some, like corals, kelp, mangroves and seagrasses, create further habitats for other organisms.

Ocean	Area million km²	%	Volume million cu km	%	Mean depth km	Max depth km	Coastline km	%
Pacific Ocean	155.6	46.4	679.6	49.6	4.37	10.924	135,663	
Atlantic Ocean	76.8	22.9	313.4	22.5	4.08	8.605	111,866	
Indian Ocean	68.6	20.4	269.3	19.6	3.93	7.258	66,526	
Southern Ocean	20.3	6.1	91.5	6.7	4.51	7.235	17,968	
Arctic Ocean	14.1	4.2	17.0	1.2	1.21	4.665	45,389	
Overall	335.3		1370.8		4.09	10.924	377,412	

The ocean occupies 71 percent of the world surface, averaging nearly four kilometres in depth. There are five major oceans, of which the Pacific Ocean is nearly as large as the rest put together. Coastlines fringe the land for nearly 380,000 kilometres.

Marine habitats can be broadly divided into pelagic and demersal habitats. Pelagic habitats are the habitats of the open water column, away from the bottom of the ocean. Demersal habitats are the habitats that are near or on the bottom of the ocean. An organism living in a pelagic habitat is said to be a pelagic organism, as in pelagic fish. Similarly, an organism living in a demersal habitat is said to be a demersal organism, as in demersal fish. Pelagic habitats are intrinsically ephemeral, depending on what ocean currents are doing.

Land runoff, pouring into the sea, can contain nutrients.

The land-based ecosystem depends on topsoil and fresh water, while the marine ecosystem depends on dissolved nutrients washed down from the land.

Ocean deoxygenation poses a threat to marine habitats, due to the growth of low oxygen zones.

Ocean Currents

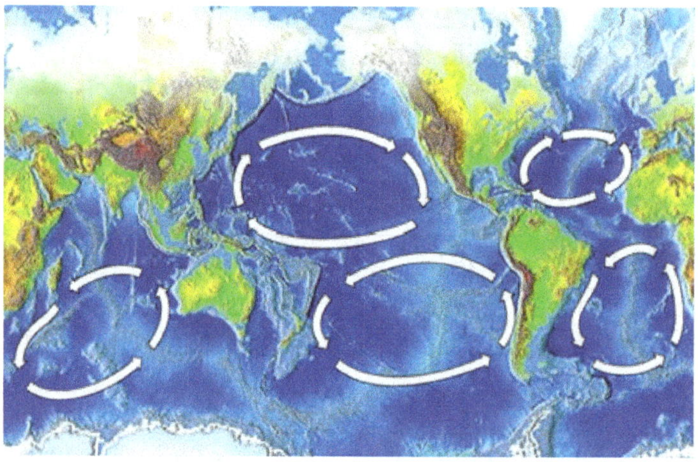

Ocean gyres rotate clockwise in the north and counterclockwise in the south.

In marine systems, ocean currents have a key role determining which areas are effective as habitats, since ocean currents transport the basic nutrients needed to support marine life. Plankton are the life forms that inhabit the ocean that are so small (less than 2 mm) that they cannot effectively propel themselves through the water, but must drift instead with the currents. If the current carries the right nutrients, and if it also flows at a suitably shallow depth where there is plenty of sunlight, then such a current itself can become a suitable habitat for photosynthesizing tiny algae called phytoplankton. These tiny plants are the primary producers in the ocean, at the start of the food chain. In turn, as the population of drifting phytoplankton grows, the water becomes a suitable habitat for zooplankton, which feed on the phytoplankton. While phytoplankton are tiny drifting plants, zooplankton are tiny drifting animals, such as the larvae of fish and marine invertebrates. If sufficient zooplankton establish themselves, the current becomes a candidate habitat for the forage fish that feed on them. And then if sufficient forage fish move to the area, it becomes a candidate habitat for larger predatory fish and other marine animals that feed on the forage fish. In this dynamic way, the current itself can, over time, become a moving habitat for multiple types of marine life.

This algae bloom occupies sunlit epipelagic waters off the southern coast of England. The algae are maybe feeding on nutrients from land runoff or upwellings at the edge of the continental shelf.

Ocean currents can be generated by differences in the density of the water. How dense water is depends on how saline or warm it is. If water contains differences in salt content or temperature, then the different densities will initiate a current. Water that is saltier or cooler will be denser, and will sink in relation to the surrounding water. Conversely, warmer and less salty water will float to the surface. Atmospheric winds and pressure differences also produces surface currents, waves and seiches. Ocean currents are also generated by the gravitational pull of the sun and moon (tides), and seismic activity (tsunami).

The rotation of the Earth affects the direction ocean currents take, and explains which way the large circular ocean gyres rotate in the image above left. Suppose a current at the equator is heading north. The Earth rotates eastward, so the water possesses that rotational momentum. But the further the water moves north, the slower the earth moves eastward. If the current could get to the North Pole, the earth wouldn't be moving eastward at all. To conserve its rotational momentum, the further the current travels north the faster it must move eastward. So the effect is that the current curves to the right. This is the Coriolis effect. It is weakest at the equator and strongest at the poles. The effect is opposite south of the equator, where currents curve left.

Marine Topography

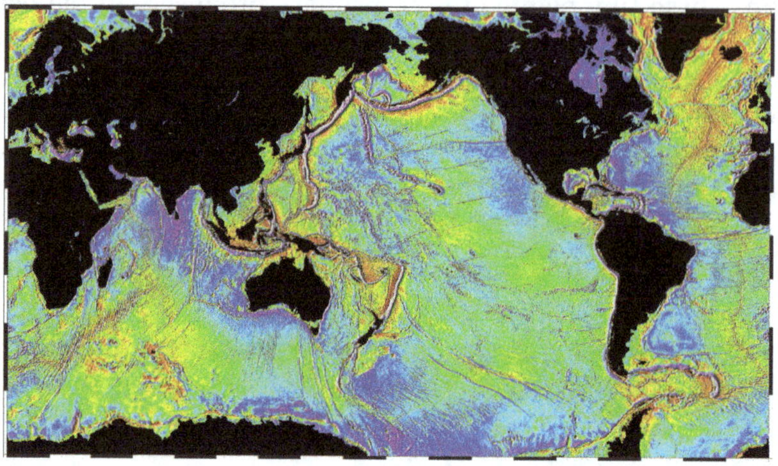

Map of underwater topography (1995 NOAA).

Marine topography refers to the shape the land has when it interfaces with the ocean. These shapes are obvious along coastlines, but they occur also in significant ways underwater. The effectiveness of marine habitats is partially defined by these shapes, including the way they interact with and shape ocean currents, and the way sunlight diminishes when these landforms occupy increasing depths.

Marine topographies include coastal and oceanic landforms ranging from coastal estuaries and shorelines to continental shelves and coral reefs. Further out in the open ocean, they include underwater and deep sea features such as ocean rises and seamounts. The submerged surface has mountainous features, including a globe-spanning mid-ocean ridge system, as well as undersea volcanoes, oceanic trenches, submarine canyons, oceanic plateaus and abyssal plains.

The mass of the oceans is approximately 1.35×10^{18} metric tons, or about 1/4400 of the total mass of the Earth. The oceans cover an area of 3.618×10^{8} km^2 with a mean depth of 3,682 m, resulting in an estimated volume of 1.332×10^{9} km^3.

Biomass

One measure of the relative importance of different marine habitats is the rate at which they produce biomass.

Producer	**Biomass productivity (gC/m²/yr)**	**Ref**	**Total area (million km²)**	**Ref**	**Total production (billion tonnes C/yr)**	**Comment**
swamps and marshes	2,500					Includes freshwater
coral reefs	2,000		0.28		0.56	
algal beds	2,000					
river estuaries	1,800					
open ocean	125		311		39	

Coastal

Coastlines can be volatile habitats.

Marine coasts are dynamic environments which constantly change, like the ocean which partially shape them. The Earth's natural processes, including weather and sea level change, result in the erosion, accretion and resculpturing of coasts as well as the flooding and creation of continental shelves and drowned river valleys.

The main agents responsible for deposition and erosion along coastlines are waves, tides and currents. The formation of coasts also depends on the nature of the rocks they are made of – the harder the rocks the less likely they are to erode, so variations in rock hardness result in coastlines with different shapes.

Tides often determine the range over which sediment is deposited or eroded. Areas with high tidal ranges allow waves to reach farther up the shore, and areas with lower tidal ranges produce deposition at a smaller elevation interval. The tidal range is influenced by the size and shape of the coastline. Tides do not typically cause erosion by themselves; however, tidal bores can erode as the waves surge up river estuaries from the ocean.

Waves erode coastline as they break on shore releasing their energy; the larger the wave the more energy it releases and the more sediment it moves. Sediment deposited by waves comes from eroded cliff faces and is moved along the coastline by the waves. Sediment deposited by rivers is the dominant influence on the amount of sediment located on a coastline.

The sedimentologist Francis Shepard classified coasts as *primary* or *secondary*.

- Primary coasts are shaped by non-marine processes, by changes in the land form. If a coast is in much the same condition as it was when sea level was stabilised after the last ice age, it is called a primary coast. "Primary coasts are created by erosion (the wearing away of soil or rock), deposition (the buildup of sediment or sand) or tectonic activity (changes in the structure of the rock and soil because of earthquakes). Many of these coastlines were formed as the sea level rose during the last 18,000 years, submerging river and glacial valleys to form bays and fjords." An example of a primary coast is a river delta, which forms when a river deposits soil and other material as it enters the sea.
- Secondary coasts are produced by marine processes, such as the action of the sea or by

creatures that live in it. Secondary coastlines include sea cliffs, barrier islands, mud flats, coral reefs, mangrove swamps and salt marshes.

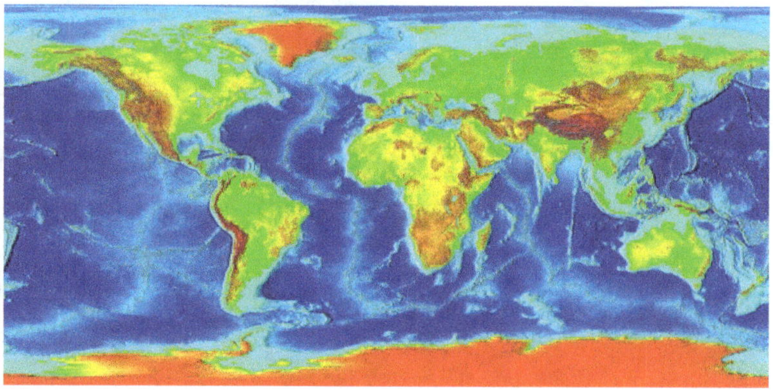

The global continental shelf, highlighted in cyan, defines the extent of coastal habitats, and occupies 5% of the total world area.

Continental coastlines usually have a continental shelf, a shelf of relatively shallow water, less than 200 metres deep, which extends 68 km on average beyond the coast. Worldwide, continental shelves occupy a total area of about 24 million km^2 (9 million sq mi), nearly 5% of the world's total area. Since the continental shelf is usually less than 200 metres deep, it follows that coastal habitats are generally photic, situated in the sunlit epipelagic zone. This means the conditions for photosynthetic processes so important for primary production, are available to coastal marine habitats. Because land is nearby, there are large discharges of nutrient rich land runoff into coastal waters. Further, periodic upwellings from the deep ocean can provide cool and nutrient rich currents along the edge of the continental shelf.

Waves and currents shape the intertidal shoreline, eroding the softer rocks and transporting and grading loose particles into shingles, sand or mud.

As a result, coastal marine life is the most abundant in the world. It is found in tidal pools, fjords and estuaries, near sandy shores and rocky coastlines, around coral reefs and on or above the continental shelf. Coastal fish include small forage fish as well as the larger predator fish that feed on them. Forage fish thrive in inshore waters where high productivity results from upwelling and shoreline run off of nutrients. Some are partial residents that spawn in streams, estuaries and bays, but most complete their life cycle in the zone. There can also be a mutualism between species that occupy adjacent marine habitats. For example, fringing reefs just below low tide level have a mutually beneficial relationship with mangrove forests at high tide level and sea grass meadows in between: the reefs protect the mangroves and seagrass from strong currents and waves that would damage them or erode the sediments in which they are rooted, while the mangroves and seagrass protect the coral from large influxes of silt, fresh water and pollutants. This additional level of variety in the environment is beneficial to many types of coral reef animals, which for example may feed in the sea grass and use the reefs for protection or breeding.

Coastal habitats are the most visible marine habitats, but they are not the only important marine habitats. Coastlines run for 380,000 kilometres, and the total volume of the ocean is 1,370 million cu km. This means that for each metre of coast, there is 3.6 cu km of ocean space available somewhere for marine habitats.

Intertidal

Intertidal zones, those areas close to shore, are constantly being exposed and covered by the ocean's tides. A huge array of life lives within this zone.

Shore habitats range from the upper intertidal zones to the area where land vegetation takes prominence. It can be underwater anywhere from daily to very infrequently. Many species here are scavengers, living off of sea life that is washed up on the shore. Many land animals also make much use of the shore and intertidal habitats. A subgroup of organisms in this habitat bores and grinds exposed rock through the process of bioerosion.

Sandy Shores

Sandy shores provide shifting homes to many species.

Sandy shores, also called beaches, are coastal shorelines where sand accumulates. Waves and currents shift the sand, continually building and eroding the shoreline. Longshore currents flow parallel to the beaches, making waves break obliquely on the sand. These currents transport large amounts of sand along coasts, forming spits, barrier islands and tombolos. Longshore currents also commonly create offshore bars, which give beaches some stability by reducing erosion.

Sandy shores are full of life, The grains of sand host diatoms, bacteria and other microscopic creatures. Some fish and turtles return to certain beaches and spawn eggs in the sand. Birds habitat beaches, like gulls, loons, sandpipers, terns and pelicans. Aquatic mammals, such sea lions, recuperate on them. Clams, periwinkles, crabs, shrimp, starfish and sea urchins are found on most beaches.

Sand is a sediment made from small grains or particles with diameters between about 60 µm and 2 mm. Mud is a sediment made from particles finer than sand. This small particle size means that mud particles tend to stick together, whereas sand particles do not. Mud is not easily shifted by waves and currents, and when it dries out, cakes into a solid. By contrast, sand is easily shifted by waves and currents, and when sand dries out it can be blown in the wind, accumulating into shifting sand dunes. Beyond the high tide mark, if the beach is low-lying, the wind can form rolling hills of sand dunes. Small dunes shift and reshape under the influence of the wind while larger dunes stabilise the sand with vegetation.

Ocean processes grade loose sediments to particle sizes other than sand, such as gravel or cobbles. Waves breaking on a beach can leave a berm, which is a raised ridge of coarser pebbles or sand, at the high tide mark. Shingle beaches are made of particles larger than sand, such as cobbles, or small stones. These beaches make poor habitats. Little life survives because the stones are churned and pounded together by waves and currents.

Rocky Shores

Tidepools on rocky shores make turbulent habitats for many forms of marine life.

The relative solidity of rocky shores seems to give them a permanence compared to the shifting nature of sandy shores. This apparent stability is not real over even quite short geological time scales, but it is real enough over the short life of an organism. In contrast to sandy shores, plants and animals can anchor themselves to the rocks.

Competition can develop for the rocky spaces. For example, barnacles can compete successfully on open intertidal rock faces to the point where the rock surface is covered with them. Barnacles resist desiccation and grip well to exposed rock faces. However, in the crevices of the same rocks, the inhabitants are different. Here mussels can be the successful species, secured to the rock with their byssal threads.

Rocky and sandy coasts are vulnerable because humans find them attractive and want to live near them. An increasing proportion of the humans live by the coast, putting pressure on coastal habitats.

Mudflats

Mudflats become temporary habitats for migrating birds.

Mudflats are coastal wetlands that form when mud is deposited by tides or rivers. They are found in sheltered areas such as bays, bayous, lagoons, and estuaries. Mudflats may be viewed geologically as exposed layers of bay mud, resulting from deposition of estuarine silts, clays and marine animal detritus. Most of the sediment within a mudflat is within the intertidal zone, and thus the flat is submerged and exposed approximately twice daily.

Mudflats are typically important regions for wildlife, supporting a large population, although levels of biodiversity are not particularly high. They are of particular importance to migratory birds. In the United Kingdom mudflats have been classified as a Biodiversity Action Plan priority habitat.

Mangrove and Salt Marshes

Mangroves provide nurseries for fish.

Mangrove swamps and salt marshes form important coastal habitats in tropical and temperate areas respectively.

Mangroves are species of shrubs and medium size trees that grow in saline coastal sediment habitats in the tropics and subtropics – mainly between latitudes 25° N and 25° S. The saline conditions tolerated by various species range from brackish water, through pure seawater (30 to 40 ppt), to water concentrated by evaporation to over twice the salinity of ocean seawater (up to 90 ppt). There are many mangrove species, not all closely related. The term "mangrove" is used generally to cover all of these species, and it can be used narrowly to cover just mangrove trees of the genus *Rhizophora*.

Mangroves form a distinct characteristic saline woodland or shrubland habitat, called a *mangrove swamp* or *mangrove forest*'. Mangrove swamps are found in depositional coastal environments, where fine sediments (often with high organic content) collect in areas protected from high-energy wave action. Mangroves dominate three quarters of tropical coastlines.

Estuaries

Estuaries occur when rivers flow into a coastal bay or inlet. They are nutrient rich and have a transition zone which moves from freshwater to saltwater.

An estuary is a partly enclosed coastal body of water with one or more rivers or streams flowing into it, and with a free connection to the open sea. Estuaries form a transition zone between river environments and ocean environments and are subject to both marine influences, such as tides, waves, and the influx of saline water; and riverine influences, such as flows of fresh water and sediment. The inflow of both seawater and freshwater provide high levels of nutrients in both the water column and sediment, making estuaries among the most productive natural habitats in the world.

Most estuaries were formed by the flooding of river-eroded or glacially scoured valleys when sea level began to rise about 10,000-12,000 years ago. They are amongst the most heavily populated areas throughout the world, with about 60% of the world's population living along estuaries and the coast. As a result, estuaries are suffering degradation by many factors, including sedimentation from soil erosion from deforestation; overgrazing and other poor farming practices; overfishing; drainage and filling of wetlands; eutrophication due to excessive nutrients from sewage and animal wastes; pollutants including heavy metals, PCBs, radionuclides and hydrocarbons from sewage inputs; and diking or damming for flood control or water diversion.

Estuaries provide habitats for a large number of organisms and support very high productivity. Estuaries provide habitats for salmon and sea trout nurseries, as well as migratory bird populations.

Two of the main characteristics of estuarine life are the variability in salinity and sedimentation. Many species of fish and invertebrates have various methods to control or conform to the shifts in salt concentrations and are termed osmoconformers and osmoregulators. Many animals also burrow to avoid predation and to live in the more stable sedimental environment. However, large numbers of bacteria are found within the sediment which have a very high oxygen demand. This reduces the levels of oxygen within the sediment often resulting in partially anoxic conditions, which can be further exacerbated by limited water flux. Phytoplankton are key primary producers in estuaries. They move with the water bodies and can be flushed in and out with the tides. Their productivity is largely dependent on the turbidity of the water. The main phytoplankton present are diatoms and dinoflagellates which are abundant in the sediment.

Kelp Forests

Kelp forests provide habitat for many marine organisms.

Kelp forests are underwater areas with a high density of kelp. They form some of the most productive and dynamic ecosystems on Earth. Smaller areas of anchored kelp are called *kelp beds*. Kelp forests occur worldwide throughout temperate and polar coastal oceans.

Kelp forests provide a unique three-dimensional habitat for marine organisms and are a source for understanding many ecological processes. Over the last century, they have been the focus of extensive research, particularly in trophic ecology, and continue to provoke important ideas that are relevant beyond this unique ecosystem. For example, kelp forests can influence coastal oceanographic patterns and provide many ecosystem services.

However, humans have contributed to kelp forest degradation. Of particular concern are the effects of overfishing nearshore ecosystems, which can release herbivores from their normal population regulation and result in the over-grazing of kelp and other algae. This can rapidly result in transitions to barren landscapes where relatively few species persist.

Frequently considered an ecosystem engineer, kelp provides a physical substrate and habitat for kelp forest communities. In algae (Kingdom: Protista), the body of an individual organism is known as a thallus rather than as a plant (Kingdom: Plantae). The morphological structure of a kelp thallus is defined by three basic structural units:

- The holdfast is a root-like mass that anchors the thallus to the sea floor, though unlike true roots it is not responsible for absorbing and delivering nutrients to the rest of the thallus;
- The stipe is analogous to a plant stalk, extending vertically from the holdfast and providing a support framework for other morphological features;
- The fronds are leaf- or blade-like attachments extending from the stipe, sometimes along its full length, and are the sites of nutrient uptake and photosynthetic activity.

In addition, many kelp species have pneumatocysts, or gas-filled bladders, usually located at the base of fronds near the stipe. These structures provide the necessary buoyancy for kelp to maintain an upright position in the water column.

The environmental factors necessary for kelp to survive include hard substrate (usually rock), high nutrients (e.g., nitrogen, phosphorus), and light (minimum annual irradiance dose > 50 E m^{-2}). Especially productive kelp forests tend to be associated with areas of significant oceanographic upwelling, a process that delivers cool nutrient-rich water from depth to the ocean's mixed surface layer. Water flow and turbulence facilitate nutrient assimilation across kelp fronds throughout the water column. Water clarity affects the depth to which sufficient light can be transmitted. In ideal conditions, giant kelp (*Macrocystis spp.*) can grow as much as 30-60 centimetres vertically per day. Some species such as *Nereocystis* are annual while others like *Eisenia* are perennial, living for more than 20 years. In perennial kelp forests, maximum growth rates occur during upwelling months (typically spring and summer) and die-backs correspond to reduced nutrient availability, shorter photoperiods and increased storm frequency.

Seagrass Meadows

White-spotted puffers like living in seagrass areas.

Seagrasses are flowering plants from one of four plant families which grow in marine environments. They are called *seagrasses* because the leaves are long and narrow and are very often green, and because the plants often grow in large meadows which look like grassland. Since seagrasses photosynthesize and are submerged, they must grow submerged in the photic zone, where there is enough sunlight. For this reason, most occur in shallow and sheltered coastal waters anchored in sand or mud bottoms.

Seagrasses form extensive beds or meadows, which can be either monospecific (made up of one species) or multispecific (where more than one species co-exist). Seagrass beds make highly diverse and productive ecosystems. They are home to phyla such as juvenile and adult fish, epiphytic and free-living macroalgae and microalgae, mollusks, bristle worms, and nematodes. Few species were originally considered to feed directly on seagrass leaves (partly because of their low nutritional content), but scientific reviews and improved working methods have shown that seagrass herbivory is a highly important link in the food chain, with hundreds of species feeding on seagrasses worldwide, including green turtles, dugongs, manatees, fish, geese, swans, sea urchins and crabs.

Seagrasses are ecosystem engineers in the sense that they partly create their own habitat. The leaves slow down water-currents increasing sedimentation, and the seagrass roots and rhizomes stabilize the seabed. Their importance to associated species is mainly due to provision of shelter (through their three-dimensional structure in the water column), and due to their extraordinarily high rate of primary production. As a result, seagrasses provide coastal zones with ecosystem services, such as fishing grounds, wave protection, oxygen production and protection against coastal erosion. Seagrass meadows account for 15% of the ocean's total carbon storage.

Coral Reefs

Reefs comprise some of the densest and most diverse habitats in the world. The best-known types of reefs are tropical coral reefs which exist in most tropical waters; however, reefs can also exist in cold water. Reefs are built up by corals and other calcium-depositing animals, usually on top of a rocky outcrop on the ocean floor. Reefs can also grow on other surfaces, which has made it possible to create artificial reefs. Coral reefs also support a huge community of life, including the corals themselves, their symbiotic zooxanthellae, tropical fish and many other organisms.

Much attention in marine biology is focused on coral reefs and the El Niño weather phenomenon. In 1998, coral reefs experienced the most severe mass bleaching events on record, when vast expanses of reefs across the world died because sea surface temperatures rose well above normal. Some reefs are recovering, but scientists say that between 50% and 70% of the world's coral reefs are now endangered and predict that global warming could exacerbate this trend.

Open Ocean

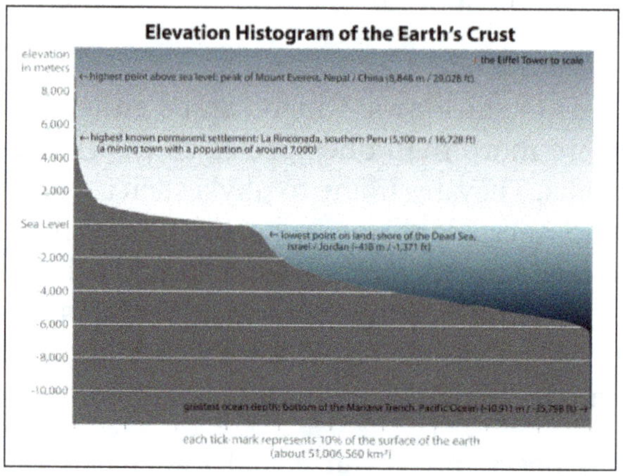

Elevation-area graph showing the proportion of land area at given heights and the proportion of ocean area at given depths.

The open ocean is relatively unproductive because of a lack of nutrients, yet because it is so vast, it has more overall primary production than any other marine habitat. Only about 10 percent of marine species live in the open ocean. But among them are the largest and fastest of all marine animals, as well as the animals that dive the deepest and migrate the longest. In the depths lurk animal that, to our eyes, appear hugely alien.

Surface Waters

In the open ocean, sunlit surface epipelagic waters get enough light for photosynthesis, but there are often not enough nutrients. As a result, large areas contain little life apart from migrating animals.

The surface waters are sunlit. The waters down to about 200 metres are said to be in the epipelagic zone. Enough sunlight enters the epipelagic zone to allow photosynthesis by phytoplankton. The epipelagic zone is usually low in nutrients. This partially because the organic debris produced in the zone, such as excrement and dead animals, sink to the depths and are lost to the upper zone. Photosynthesis can happen only if both sunlight and nutrients are present.

In some places, like at the edge of continental shelves, nutrients can upwell from the ocean depth, or land runoff can be distributed by storms and ocean currents. In these areas, given that both sunlight and nutrients are now present, phytoplankton can rapidly establish itself, multiplying so fast that the water turns green from the chlorophyll, resulting in an algal bloom. These nutrient rich surface waters are among the most biologically productive in the world, supporting billions of tonnes of biomass.

"Phytoplankton are eaten by zooplankton - small animals which, like phytoplankton, drift in the ocean currents. The most abundant zooplankton species are copepods and krill: tiny crustaceans that are the most numerous animals on Earth. Other types of zooplankton include jelly fish and the larvae of fish, marine worms, starfish, and other marine organisms". In turn, the zooplankton are eaten by filter feeding animals, including some seabirds, small forage fish like herrings and sardines, whale sharks, manta rays, and the largest animal in the world, the blue whale. Yet again, moving up the foodchain, the small forage fish are in turn eaten by larger predators, such as tuna, marlin, sharks, large squid, seabirds, dolphins, and toothed whales.

Deep Sea

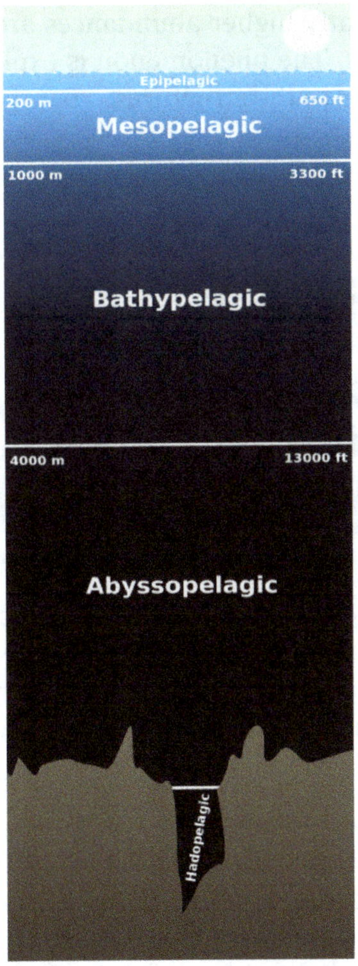

Scale diagram of the layers of the pelagic zone.

The deep sea starts at the aphotic zone, the point where sunlight loses most of its energy in the water. Many life forms that live at these depths have the ability to create their own light a unique evolution known as bio-luminescence.

In the deep ocean, the waters extend far below the epipelagic zone, and support very different types of pelagic life forms adapted to living in these deeper zones.

Much of the aphotic zone's energy is supplied by the open ocean in the form of detritus. In deep water, marine snow is a continuous shower of mostly organic detritus falling from the upper layers of the water column. Its origin lies in activities within the productive photic zone. Marine snow includes dead or dying plankton, protists (diatoms), fecal matter, sand, soot and other inorganic dust. The "snowflakes" grow over time and may reach several centimetres in diameter, travelling for weeks before reaching the ocean floor. However, most organic components of marine snow are consumed by microbes, zooplankton and other filter-feeding animals within the first 1,000 metres of their journey, that is, within the epipelagic zone. In this way marine snow may be considered the foundation of deep-sea mesopelagic and benthic ecosystems: As sunlight cannot reach them, deep-sea organisms rely heavily on marine snow as an energy source.

Some deep-sea pelagic groups, such as the lanternfish, ridgehead, marine hatchetfish, and lightfish families are sometimes termed *pseudoceanic* because, rather than having an even distribution in open water, they occur in significantly higher abundances around structural oases, notably seamounts and over continental slopes. The phenomenon is explained by the likewise abundance of prey species which are also attracted to the structures.

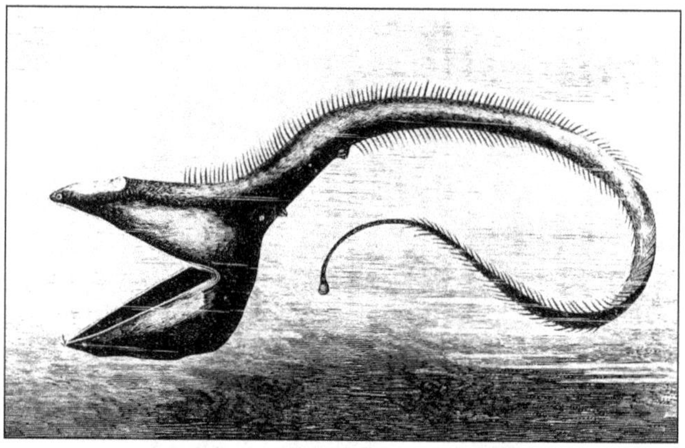

The umbrella mouth gulper eel can swallow a fish much larger than itself.

The fish in the different pelagic and deep water benthic zones are physically structured, and behave in ways, that differ markedly from each other. Groups of coexisting species within each zone all seem to operate in similar ways, such as the small mesopelagic vertically migrating plankton-feeders, the bathypelagic anglerfishes, and the deep water benthic rattails. "

Ray finned species, with spiny fins, are rare among deep sea fishes, which suggests that deep sea fish are ancient and so well adapted to their environment that invasions by more modern fishes have been unsuccessful. The few ray fins that do exist are mainly in the Beryciformes and Lampriformes, which are also ancient forms. Most deep sea pelagic fishes belong to their own orders, suggesting a long evolution in deep sea environments. In contrast, deep water benthic species, are in orders that include many related shallow water fishes.

The umbrella mouth gulper is a deep sea eel with an enormous loosely hinged mouth. It can open its mouth wide enough to swallow a fish much larger than itself, and then expand its stomach to accommodate its catch.

Sea Floor

Vents and Seeps

Zooarium chimney provides a habitat for vent biota.

Hydrothermal vents along the mid-ocean ridge spreading centers act as oases, as do their opposites, cold seeps. Such places support unique biomes and many new microbes and other lifeforms have been discovered at these locations.

Trenches

The deepest recorded oceanic trenches measure to date is the Mariana Trench, near the Philippines, in the Pacific Ocean at 10,924 m (35,838 ft). At such depths, water pressure is extreme and there is no sunlight, but some life still exists. A white flatfish, a shrimp and a jellyfish were seen by the American crew of the bathyscaphe *Trieste* when it dove to the bottom in 1960.

Seamounts

Marine life also flourishes around seamounts that rise from the depths, where fish and other sea life congregate to spawn and feed.

Classification of Marine Habitats

Coast

Rugged coastline of the West Coast Region of New Zealand.

A coastline or a seashore is the area where land meets the sea or ocean, or a line that forms the boundary between the land and the ocean or a lake. A precise line that can be called a coastline cannot be determined due to the Coastline paradox.

The term *coastal zone* is a region where interaction of the sea and land processes occurs. Both the terms coast and coastal are often used to describe a geographic location or region; for example, New Zealand's West Coast, or the East and West Coasts of the United States. Edinburgh for example is a city on the coast of Scotland.

A pelagic coast refers to a coast which fronts the open ocean, as opposed to a more sheltered coast in a gulf or bay. A shore, on the other hand, can refer to parts of the land which adjoin any large body of water, including oceans (sea shore) and lakes (lake shore). Similarly, the somewhat related term "[Stream bed/bank]" refers to the land alongside or sloping down to a river (riverbank) or to a body of water smaller than a lake. "Bank" is also used in some parts of the world to refer to an artificial ridge of earth intended to retain the water of a river or pond; in other places this may be called a levee.

While many scientific experts might agree on a common definition of the term "coast", the delineation of the extents of a coast differ according to jurisdiction, with many scientific and government authorities in various countries differing for economic and social policy reasons. According to the UN atlas, 44% of people live within 150 kilometres (93 miles) of the sea.

Formation

Atlantic rocky coastline, showing a surf area. Porto Covo, west coast of Portugal.

Tides often determine the range over which sediment is deposited or eroded. Areas with high tidal ranges allow waves to reach farther up the shore, and areas with lower tidal ranges produce deprossosition at a smaller elevation interval. The tidal range is influenced by the size and shape of the coastline. Tides do not typically cause erosion by themselves; however, tidal bores can erode as the waves surge up river estuaries from the ocean.

Waves erode coastline as they break on shore releasing their energy; the larger the wave the more energy it releases and the more sediment it moves. Coastlines with longer shores have more room for the waves to disperse their energy, while coasts with cliffs and short shore faces give little room for the wave energy to be dispersed. In these areas the wave energy breaking against the cliffs is higher, and air and water are compressed into cracks in the rock, forcing the rock apart, breaking it down. Sediment deposited by waves comes from eroded cliff faces and is moved along the coastline by the waves. This forms an abrasion or cliffed coast.

Sediment deposited by rivers is the dominant influence on the amount of sediment located on a coastline. Today riverine deposition at the coast is often blocked by dams and other human regulatory devices, which remove the sediment from the stream by causing it to be deposited inland.

Like the ocean which shapes them, coasts are a dynamic environment with constant change. The Earth's natural processes, particularly sea level rises, waves and various weather phenomena, have resulted in the erosion, accretion and reshaping of coasts as well as flooding and creation of continental shelves and drowned river valleys (rias).

Environmental Importance

Somalia has the longest coastline in Africa.

The coast and its adjacent areas on and off shore are an important part of a local ecosystem: the mixture of fresh water and salt water (brackish water) in estuaries provides many nutrients for marine life. Salt marshes and beaches also support a diversity of plants, animals and insects crucial to the food chain.

The high level of biodiversity creates a high level of biological activity, which has attracted human activity for thousands of years.

Human Impacts

Human Uses of Coasts

This stretch of coast in Tanzania's capital Dar es Salaam serves as a public waste dump.

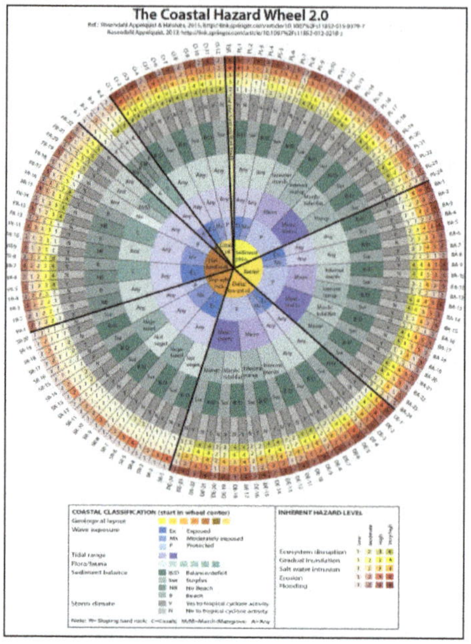

The Coastal Hazard Wheel system published by UNEP for global coastal management.

A settled coastline in Marblehead, Massachusetts. Once a fishing port, the harbor is now dedicated to tourism and pleasure boating. Observe that the sand and rocks have been darkened by oil slick up to the high-water line.

The coast is a frontier that nations have typically defended against military invaders, smugglers and illegal migrants. Fixed coastal defenses have long been erected in many nations and coastal countries typically have a navy and some form of coast guard.

Coasts, especially those with beaches and warm water, attract tourists. In many island nations such as those of the Mediterranean, South Pacific and Caribbean, tourism is central to the economy. Coasts offer recreational activities such as swimming, fishing, surfing, boating, and sunbathing. Growth management can be a challenge for coastal local authorities who often struggle to provide the infrastructure required by new residents.

Threats to a Coast

Coasts also face many human-induced environmental impacts. The human influence on climate change is thought to contribute to an accelerated trend in sea level rise which threatens coastal habitats.

Pollution can occur from a number of sources: garbage and industrial debris; the transportation of petroleum in tankers, increasing the probability of large oil spills; small oil spills created by large and small vessels, which flush bilge water into the ocean.

Fishing has declined due to habitat degradation, overfishing, trawling, bycatch and climate change. Since the growth of global fishing enterprises after the 1950s, intensive fishing has spread from a few concentrated areas to encompass nearly all fisheries. The scraping of the ocean floor in bottom dragging is devastating to coral, sponges and other long-lived species that do not recover quickly. This destruction alters the functioning of the ecosystem and can permanently alter species composition and biodiversity. Bycatch, the capture of unintended species in the course of fishing, is typically returned to the ocean only to die from injuries or exposure. Bycatch represents about a quarter of all marine catch. In the case of shrimp capture, the bycatch is five times larger than the shrimp caught.

It is believed that melting Arctic ice will cause sea levels to rise and flood costal areas.

Conservation

Extraordinary population growth in the 21st century has placed stress on the planet's ecosystems. For example, on Saint Lucia, harvesting mangrove for timber and clearing for fishing reduced the mangrove forests, resulting in a loss of habitat and spawning grounds for marine life that was unique to the area. These forests also helped to stabilize the coastline. Conservation efforts since the 1980s have partially restored the ecosystem.

Types of Coast

According to one principle of classification, an emergent coastline is a coastline which has experienced a fall in sea level, because of either a global sea level change, or local uplift. Emergent coastlines are identifiable by the coastal landforms, which are above the high tide mark, such as raised beaches. In contrast, a submergent coastline is one where the sea level has risen, due to a global sea level change, local subsidence, or isostatic rebound. Submergent coastlines are identifiable by their submerged, or "drowned" landforms, such as rias (drowned valleys) and fjords.

According to a second principle of classification, a concordant coastline is a coastline where bands of different rock types run parallel to the shore. These rock types are usually of varying resistance, so the coastline forms distinctive landforms, such as coves. Discordant coastlines feature distinctive landforms because the rocks are eroded by ocean waves. The less resistant rocks erode faster, creating inlets or bay; the more resistant rocks erode more slowly, remaining as headlands or outcroppings.

Other coastal categories:

- A cliffed coast or abrasion coast is one where marine action has produced steep declivities known as cliffs.
- A flat coast is one where the land gradually descends into the sea.
- A graded shoreline is one where wind and water action has produced a flat and straight coastline.

Coastal Landforms

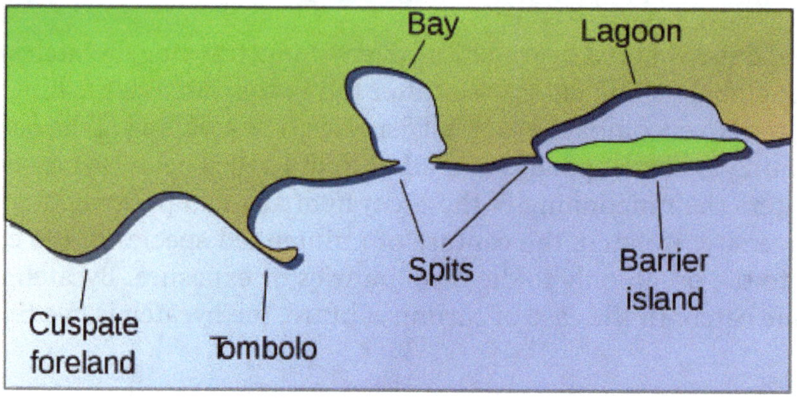

Coastal landforms. The feature shown here as a bay would, in certain (mainly southern) parts of Britain, be called a cove. That between the cuspate foreland and the tombolo is a British bay.

The following articles describe some coastal landforms.

• Bay	• Cove
• Headland	• Peninsula

Cliff Erosion

- Much of the sediment deposited along a coast is the result of erosion of a surrounding cliff, or bluff. Sea cliffs retreat landward because of the constant undercutting of slopes by waves. If the slope/cliff being undercut is made of unconsolidated sediment it will erode at a much faster rate then a cliff made of bedrock.
- A natural arch is formed when a sea stack is eroded through by waves.
- Sea caves are made when certain rock beds are more susceptible to erosion than the surrounding rock beds because of different areas of weakness. These areas are eroded at a faster pace creating a hole or crevasse that, through time, by means of wave action and erosion, becomes a cave.
- A stack is formed when a headland is eroded away by wave and wind action.
- A stump is a shortened sea stack that has been eroded away or fallen because of instability.
- Wave-cut notches are caused by the undercutting of overhanging slopes which leads to increased stress on cliff material and a greater probability that the slope material will fall. The fallen debris accumulates at the bottom of the cliff and is eventually removed by waves.
- A wave-cut platform forms after erosion and retreat of a sea cliff has been occurring for a long time. Gently sloping wave-cut platforms develop early on in the first stages of cliff retreat. Later the length of the platform decreases because the waves lose their energy as they break further off shore.

Coastal Features Formed by Sediment

• Beach • Beach cusps	• Cuspate foreland • Dune system • Mudflat • Raised beach	• Ria • Shoal • Strand plain	• Spit • Surge channel • Tombolo

Coastal Features Formed by Another Feature

- Lagoon
- Salt marsh

Other Features on the Coast

• Concordant coastline • Discordant coastline	• Fjord • Island	• Island arc • Machair

Coastal Processes

The following articles describe the various geologic processes that affect a coastal zone:

• Attrition • Currents • Denudation • Deposition • Erosion • Flooding	• Longshore drift • Saltation • Sea level change o eustatic o isostatic	• Sedimentation • Coastal sediment supply o sediment transport o solution o subaerial processes o suspen-sion	• Tides • Water waves o diffraction o refraction o wave break-ing o wave shoaling • Weathering

Wildlife

Animals

Some of the animals live along a typical coast. There are animals like puffins, sea turtles and rock-hopper penguins. Sea snails and various kinds of barnacles live on the coast and scavenge on food deposited by the sea. Most coastal animals are used to humans in developed areas, such as dolphins and seagulls who eat food thrown for them by tourists. Since the coastal areas are all part of the littoral zone, there is a profusion of marine life found just off-coast.

There are many kinds of seabirds on the coast. Pelicans and cormorants join up with terns and oystercatchers to forage for fish and shellfish on the coast. There are sea lions on the coast of Wales and other countries.

Plants

Coastal areas are famous for their kelp beds. Kelp is a fast-growing seaweed that grows up to a metre a day. Corals and sea anemones are true animals, but live a lifestyle similar to that of plants. Mangroves, seagrasses and salt marsh are important coastal vegetation types in tropical and temperate environments respectively.

Coastline Statistics

Coastline Problem

Shortly before 1951, Lewis Fry Richardson, in researching the possible effect of border lengths on the probability of war, noticed that the Portuguese reported their measured border with Spain to be 987 km, but the Spanish reported it as 1214 km. This was the beginning of the coastline problem, which is a mathematical uncertainty inherent in the measurement of boundaries that are irregular.

The prevailing method of estimating the length of a border (or coastline) was to lay out n equal straight-line segments of length ℓ with dividers on a map or aerial photograph. Each end of the segment must be on the boundary. Investigating the discrepancies in border estimation, Richardson discovered what is now termed the Richardson Effect: the sum of the segments is inversely proportional to the common length of the segments. In effect, the shorter the ruler, the longer the measured border; the Spanish and Portuguese geographers were simply using different-length rulers.

The result most astounding to Richardson is that, under certain circumstances, as ℓ approaches zero, the length of the coastline approaches infinity. Richardson had believed, based on Euclidean geometry, that a coastline would approach a fixed length, as do similar estimations of regular geometric figures. For example, the perimeter of a regular polygon inscribed in a circle approaches the circumference with increasing numbers of sides (and decrease in the length of one side). In geometric measure theory such a smooth curve as the circle that can be approximated by small straight segments with a definite limit is termed a rectifiable curve.

Measuring a Coastline

More than a decade after Richardson completed his work, Benoit Mandelbrot developed a new branch of mathematics, fractal geometry, to describe just such non-rectifiable complexes in nature as the infinite coastline. His own definition of the new figure serving as the basis for his study is:

I coined *fractal* from the Latin adjective *fractus*. The corresponding Latin verb *frangere* means "to break:" to create irregular fragments. It is therefore sensible ... that, in addition to "fragmented" ... *fractus* should also mean "irregular."

A key property of the fractal is self-similarity; that is, at any scale the same general configuration appears. A coastline is perceived as bays alternating with promontories. In the hypothetical situation that a given coastline has this property of self-similarity, then no matter how greatly any one small section of coastline is magnified, a similar pattern of smaller bays and promontories superimposed on larger bays and promontories appears, right down to the grains of sand. At that scale the coastline appears as a momentarily shifting, potentially infinitely long thread with a stochastic arrangement of bays and promontories formed from the small objects at hand. In such an environment (as opposed to smooth curves) Mandelbrot asserts "coastline length turns out to be an elusive notion that slips between the fingers of those who want to grasp it."

There are different kinds of fractals. A coastline with the stated property is in "a first category of fractals, namely curves whose fractal dimension is greater than 1." That last statement represents an extension by Mandelbrot of Richardson's thought. Mandelbrot's statement of the Richardson Effect is:

$$L(\epsilon) \sim F\epsilon^{1-D}$$

where L, coastline length, a function of the measurement unit, ε, is approximated by the expression. F is a constant and D is a parameter that Richardson found depended on the coastline approximated by L. He gave no theoretical explanation but Mandelbrot identified D with a non-integer form of the Hausdorff dimension, later the fractal dimension. Rearranging the right side of the expression obtains:

$$\frac{F}{\grave{o}^{D}} \cdot \grave{o}$$

where $F\varepsilon^{-D}$ must be the number of units ε required to obtain L. The fractal dimension is the number of the dimensions of the figure being used to approximate the fractal: 0 for a dot, 1 for a line, 2 for a square. D in the expression is between 1 and 2, for coastlines typically less than 1.5. The broken line measuring the coast does not extend in one direction nor does it represent an area, but is intermediate. It can be interpreted as a thick line or band of width 2ε. More broken coastlines have greater D and therefore L is longer for the same ε. Mandelbrot showed that D is independent of ε.

Pelagic Zone

Any water in a sea or lake that is neither close to the bottom nor near the shore can be said to be in the pelagic zone. The word "pelagic" is derived from Greek *(pélagos)*, meaning "open sea". The pelagic zone can be thought of in terms of an imaginary cylinder or water column that goes from the surface of the sea almost to the bottom. Conditions differ deeper in the water column such that as pressure increases with depth, the temperature drops and less light penetrates. Depending on the depth, the water column, rather like the Earth's atmosphere, may be divided into different layers.

The pelagic zone occupies 1,330 million km^3 (320 million mi^3) with a mean depth of 3.68 km (2.29 mi) and maximum depth of 11 km (6.8 mi). Fish that live in the pelagic zone are called pelagic fish. Pelagic life decreases with increasing depth. It is affected by light intensity, pressure, temperature, salinity, the supply of dissolved oxygen and nutrients, and the submarine topography, which is called bathymetry. In deep water, the pelagic zone is sometimes called the open-ocean zone and can be contrasted with water that is near the coast or on the continental shelf. In other contexts, coastal water not near the bottom is still said to be in the pelagic zone.

The pelagic zone can be contrasted with the benthic and demersal zones at the bottom of the sea. The benthic zone is the ecological region at the very bottom of the sea. It includes the sediment surface and some subsurface layers. Marine organisms living in this zone, such as clams and crabs, are called benthos. The demersal zone is just above the benthic zone. It can be significantly affected by the seabed and the life that lives there. Fish that live in the demersal zone are called demersal fish, which can be divided into benthic fish, which are denser than water so they can rest on the bottom, and benthopelagic fish, which swim in the water column just above the bottom. Demersal fish are also known as bottom feeders and groundfish.

Depth and Layers

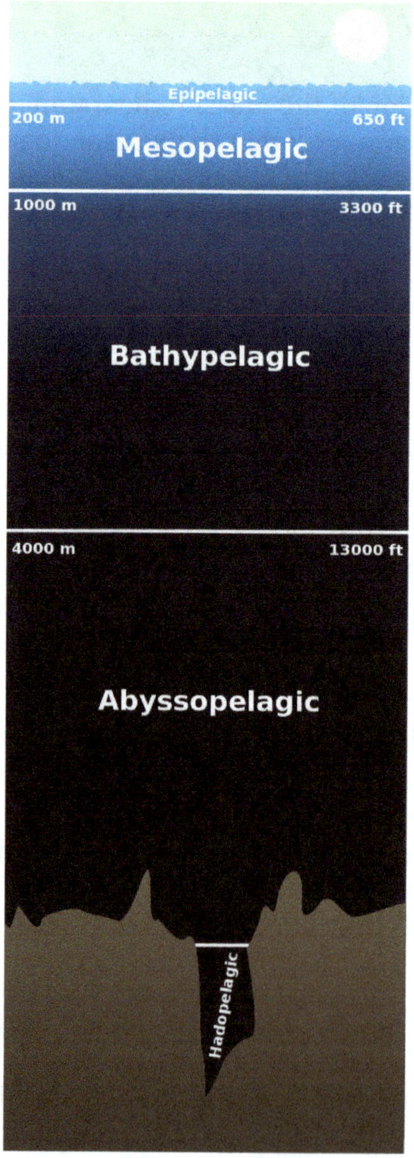

A scale diagram of the layers of the pelagic zone.

Depending on how deep the sea is, the pelagic zone can extend over up to five horizontal layers in the ocean. From the top down, these are:

Epipelagic (Sunlight)

This is the illuminated zone at the surface of the sea where enough light is available for photosynthesis. Nearly all primary production in the ocean occurs here. Consequently, plants and animals are largely concentrated in this zone. Examples of organisms living in this zone are plankton, floating seaweed, jellyfish, tuna, many sharks and dolphins.

Mesopelagic (Twilight)

The name for this zone stems from Greek μέσον *(meson)*, meaning "middle". Although some light penetrates this second layer it is insufficient for photosynthesis. At about 500 m the water also becomes depleted of oxygen. Organisms survive this environment by having more efficient gills or by minimizing movement.

Examples of animals that live here are swordfish, squid, wolffish and some species of cuttlefish. Many organisms that live in this zone are bioluminescent. Some creatures living in the mesopelagic zone rise to the epipelagic zone at night to feed.

Bathypelagic (Midnight)

The name stems from Greek *(bathýs)*, meaning "deep". At this depth, the ocean is pitch black, apart from occasional bioluminescent organisms, such as lanternfish. No living plant life exists here. Most animals living here survive by consuming the detritus falling from the zones above, which is known as "marine snow", or, like the marine hatchetfish, by preying on other inhabitants of this zone. Other examples of this zone's inhabitants are giant squid, smaller squids and the dumbo octopus. The giant squid is hunted here by deep-diving sperm whales.

Abysso pelagic (Lower Midnight)

The name is derived from Greek *(ábyssos)*, meaning "bottomless" (a holdover from the times when the deep ocean, or abyss, was believed to be bottomless). Very few creatures live in the cold temperatures, high pressures and complete darkness of this depth. Among the species found in this zone are several species of squid; echinoderms including the basket star, swimming cucumber, and the sea pig; and marine arthropods including the sea spider. Many of the species living at these depths are transparent and eyeless because of the total lack of light in this zone.

Hadopelagic

The Deep Water in Ocean Trenches

The name is derived from the realm of Hades, the underworld in Greek mythology. This zone is mostly unknown, and very few species are known to live here (in the open areas). However, many organisms live in hydrothermal vents in this and other zones. Some define the hadopelagic as waters below 6,000 m (20,000 ft), whether in a trench or not. The bathypelagic, abyssopelagic, and hadopelagic zones are very similar in character, and some marine biologists combine them into a single zone or consider the latter two to be the same. The abyssal plain is covered with soft sludge composed of dead organisms from above.

Pelagic Ecosystem

The pelagic sooty tern spends months at a time flying at sea, returning to land only for breeding.

The pelagic ecosystem is based on the phytoplankton which occupy the start of the foodchain. Phytoplankton manufacture their own food using a process of photosynthesis. Because they need sunlight, they inhabit the upper, sunlit epipelagic zone, which includes the coastal or neritic zone. Biodiversity diminishes markedly in the deeper zones below the epipelagic zone as dissolved oxygen diminishes, water pressure increases, temperatures become colder, food sources become scarce, and light diminishes and finally disappears.

Pelagic Birds

Pelagic birds, also called oceanic birds, live on the open sea, rather than around waters adjacent to land or around inland waters. Pelagic birds feed on planktonic crustaceans, squid and forage fish. Examples are the Atlantic puffin, macaroni penguins, sooty terns, shearwaters, and procellariiforms such as the albatross, procellariids and petrels.

The term seabird includes birds which live around the sea adjacent to land, as well as pelagic birds.

Pelagic Fish

Pelagic fish live in the water column of coastal, ocean, and lake waters, but not on or near the bottom of the sea or the lake. They can be contrasted with demersal fish, which live on or near the bottom, and reef fish, which are associated with coral reefs.

These fish are often migratory forage fish, which feed on plankton, and the larger fish that follow and feed on the forage fish. Examples of migratory forage fish are herring, anchovies, capelin, and menhaden. Examples of larger pelagic fish which prey on the forage fish are billfish, tuna, and oceanic sharks.

Pelagic Invertebrates

Some examples of pelagic invertebrates include krill, copepods, jellyfish, decapod larvae, hyperiid amphipods, rotifers and cladocerans.

Thorson's rule states that benthic marine invertebrates at low latitudes tend to produce large numbers of eggs developing to widely dispersing pelagic larvae, whereas at high latitudes such organisms tend to produce fewer and larger lecithotrophic (yolk-feeding) eggs and larger offspring.

Pelagic Reptiles

Pelamis platura, the pelagic sea snake, is the only one of the 65 species of marine snakes to spend its entire life in the pelagic zone. It bears live young at sea and is helpless on land. The species sometimes forms aggregations of thousands along slicks in surface waters. The pelagic sea snake is the world's most widely distributed snake species.

Many species of sea turtles spend the first years of their lives in the pelagic zone, moving closer to shore as they reach maturity.

Demersal Zone

The demersal zone is the part of the sea or ocean (or deep lake) comprising the water column that is near to (and is significantly affected by) the seabed and the benthos. The demersal zone is just above the benthic zone and forms a layer of the larger profundal zone.

Being just above the ocean floor, the demersal zone is variable in depth and can be part of the photic zone where light can penetrate and photosynthetic organisms grow, or the aphotic zone, which begins between depths of roughly 200 and 1,000 m (700 and 3,300 ft) and extends to the ocean depths, where no light penetrates.

Fish

The distinction between demersal species of fish and pelagic species is not always clear cut. The Atlantic cod (*Gadus morhua*) is a typical demersal fish but can also be found in the open water column, and the Atlantic herring (*Clupea harengus*) is predominantly a pelagic species but forms large aggregations near the seabed when it spawns on banks of gravel.

Two types of fish inhabit the demersal zone, those that are heavier than water and rest on the seabed, and those that have neutral buoyancy and remain just above the substrate. In many species of fish, neutral buoyancy is maintained by a gas-filled swim bladder which can be expanded or contracted as the circumstances require. A disadvantage of this method is that adjustments need to be made constantly as the water pressure varies when the fish swims higher and lower in the water column. An alternative buoyancy aid is the use of lipids; these are less dense than water and squalene, commonly found in shark livers, has a specific gravity of just 0.86. In the velvet belly lanternshark (*Etmopterus spinax*), a benthopelagic species, 17% of the bodyweight is liver and 70% is lipids. Benthic rays and skates have smaller livers with lower concentrations of lipids; they are therefore denser than water and they do not swim continuously, intermittently resting on the seabed. Some fish have no buoyancy aids but use their pectoral fins which are so angled as to give lift as they swim. The disadvantage of this is that, if they stop swimming, the fish sink, and they cannot hover, or swim backwards.

Demersal fish have various feeding strategies; many feed on zooplankton or organisms or algae on the seabed; some of these feed on epifauna (invertebrates on top of the seafloor), while others specialise on infauna (invertebrates that burrow beneath the seafloor). Others are scavengers, eating the dead remains of plants or animals while still others are predators.

Invertebrates

Zooplankton are animals that drift with the current, but many have some limited means of locomotion and have some control over the depths at which they drift. They use gas-filled sacs or accumulations of substances with low densities to provide buoyancy, or they may have structures that slow down any passive descent. Where the adult, benthic organism is limited to life in a certain range of depths, their larvae need to optimise their chances of settling on a suitable substrate.

Cuttlefish are able to adjust their buoyancy using their cuttlebones, lightweight rigid structures with cavities filled with gas, which have a specific gravity of about 0.6. This enables them to swim at varying depths. Another invertebrate that feeds on the seabed and has swimming abilities is the nautilus, which stores gas in its chambers and adjusts its buoyancy by use of osmosis, pumping water in and out.

Chapter 5

Corals and Coral Reef

Corals are invertebrate animals belonging to the phylum Cnidaria and can be classified as either hard or soft corals. Hard corals absorb the calcium from seawater to form hardened carbonate structures which have developed into large structures called coral reefs, providing home to a large number of species. This chapter provides a detailed and close-up analysis of corals and the reefs they have help build.

Corals

Corals are marine invertebrates in the class Anthozoa of phylum Cnidaria. They typically live in compact colonies of many identical individual polyps. The group includes the important reef builders that inhabit tropical oceans and secrete calcium carbonate to form a hard skeleton.

A coral "group" is a colony of myriad genetically identical polyps. Each polyp is a sac-like animal typically only a few millimeters in diameter and a few centimeters in length. A set of tentacles surround a central mouth opening. An exoskeleton is excreted near the base. Over many generations, the colony thus creates a large skeleton that is characteristic of the species. Individual heads grow by asexual reproduction of polyps. Corals also breed sexually by spawning: polyps of the same species release gametes simultaneously over a period of one to several nights around a full moon.

Although some corals can catch small fish and plankton, using stinging cells on their tentacles, most corals obtain the majority of their energy and nutrients from photosynthetic unicellular dinoflagellates in the genus *Symbiodinium* that live within their tissues. These are commonly known as zooxanthellae and the corals that contain them are zooxanthellate corals. Such corals require sunlight and grow in clear, shallow water, typically at depths shallower than 60 metres (200 ft). Corals are major contributors to the physical structure of the coral reefs that develop in tropical and subtropical waters, such as the enormous Great Barrier Reef off the coast of Queensland, Australia.

Other corals do not rely on zooxanthellae and can live in much deeper water, with the cold-water genus *Lophelia* surviving as deep as 3,000 metres (9,800 ft). Some have been found on the Darwin Mounds, north-west of Cape Wrath, Scotland. Corals have also been found as far north as off the coast of Washington State and the Aleutian Islands.

Taxonomy

In his *Scala Naturae*, Aristotle classified corals as "zoophyta" ("plant-animals"), animals that had characteristics of plants and were therefore hypothetically in between animals and plants. The Persian polymath Al-Biruni (d. 1048) classified sponges and corals as animals, arguing that they respond to touch. Nevertheless, people believed corals to be plants until the eighteenth century, when William Herschel used a microscope to establish that coral had the characteristic thin cell membranes of an animal.

The phylogeny of Anthozoans is not clearly understood and a number of different models have been proposed. Within the Hexacorallia, the sea anemones, coral anemones and stony corals may constitute a monophyletic grouping united by their eight-fold symmetry and cnidocyte trait. The Octocorallia appears to be monophyletic, and primitive members of this group may have been stolonate. The cladogram presented here comes from a 2014 study by Stampar et al. which was based on the divergence of mitochondrial DNA within the group and on nuclear markers.

Corals are classified in the class Anthozoa of the phylum Cnidaria. They are divided into three subclasses, Hexacorallia, Octocorallia, and Ceriantharia. The Hexacorallia include the stony corals, the sea anemones and the zoanthids. These groups have polyps that generally have 6-fold symmetry. The Octocorallia include blue coral, soft corals, sea pens, and gorgonians (sea fans and sea whips). These groups have polyps with 8-fold symmetry, each polyp having eight tentacles and eight mesenteries. Ceriantharia are the tube-dwelling anemones.

Fire corals are not true corals, being in the order Anthomedusa (sometimes known as Anthoathecata) of the class Hydrozoa.

Anatomy

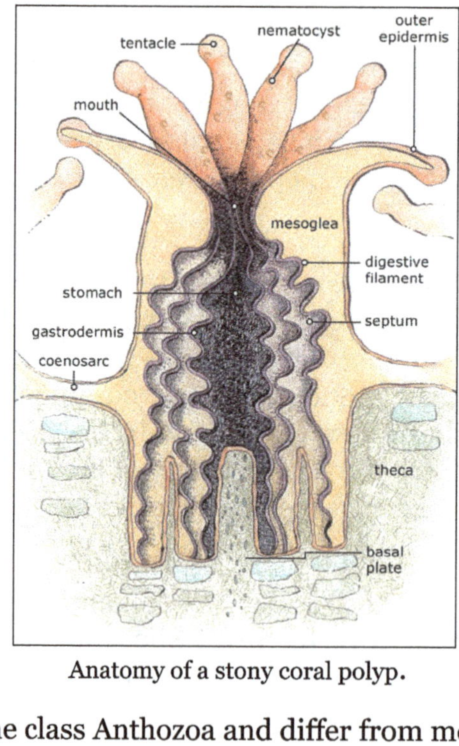

Anatomy of a stony coral polyp.

Corals are sessile animals in the class Anthozoa and differ from most other cnidarians in not having a medusa stage in their life cycle. The body unit of the animal is a polyp. Most corals are colonial, the initial polyp budding to produce another and the colony gradually developing from this small start. In stony corals, also known as hard corals, the polyps produce a skeleton composed of calcium carbonate to strengthen and protect the organism. This is deposited by the polyps and by the coenosarc, the living tissue that connects them. The polyps sit in cup-shaped depressions in the skeleton known as corallites. Colonies of stony coral are very variable in appearance; a single species may adopt an encrusting, plate-like, bushy, columnar or massive solid structure, the various forms often being linked to different types of habitat, with variations in light level and water movement being significant.

In soft corals, there is no stony skeleton but the tissues are often toughened by the presence of tiny skeletal elements known as sclerites, which are made from calcium carbonate. Soft corals are very variable in form and most are colonial. A few soft corals are stolonate, but the polyps of most are connected by sheets of coenosarc. In some species this is thick and the polyps are deeply embedded. Some soft corals are encrusting or form lobes. Others are tree-like or whip-like and have a central axial skeleton embedded in the tissue matrix. This is composed either of a fibrous protein called gorgonin or of a calcified material. In both stony and soft corals, the polyps can be retracted, with stony corals relying on their hard skeleton and cnidocytes for defence against predators, with soft corals generally relying on chemical defences in the form of toxic substances present in the tissues known as terpenoids.

Montastraea cavernosa polyps with tentacles extended.

The polyps of stony corals have six-fold symmetry while those of soft corals have eight. The mouth of each polyp is surrounded by a ring of tentacles. In stony corals these are cylindrical and taper to a point, but in soft corals they are pinnate with side branches known as pinnules. In some tropical species these are reduced to mere stubs and in some they are fused to give a paddle-like appearance. In most corals, the tentacles are retracted by day and spread out at night to catch plankton and other small organisms. Shallow water species of both stony and soft corals can be zooxanthellate, the corals supplementing their plankton diet with the products of photosynthesis produced by these symbionts. The polyps interconnect by a complex and well-developed system of gastrovascular canals, allowing significant sharing of nutrients and symbionts.

Ecology

Feeding

Polyps feed on a variety of small organisms, from microscopic zooplankton to small fish. The polyp's tentacles immobilize or kill prey using their nematocysts. These cells carry venom which they rapidly release in response to contact with another organism. A dormant nematocyst discharges in response to nearby prey touching the trigger (cnidocil). A flap (operculum) opens and its stinging apparatus fires the barb into the prey. The venom is injected through the hollow filament to immobilise the prey; the tentacles then manoeuvre the prey to the mouth.

The tentacles then contract to bring the prey into the stomach. Once the prey is digested, the stomach reopens, allowing the elimination of waste products and the beginning of the next hunting cycle. They can scavenge drifting organic molecules and dissolved organic molecules.

Intracellular Symbionts

Many corals, as well as other cnidarian groups such as *Aiptasia* (a sea anemone) form a symbiotic relationship with a class of dinoflagellate algae, zooxanthellae of the genus *Symbiodinium*. *Aiptasia*, a familiar pest among coral reef aquarium hobbyists, serves as a valuable model organism in the study of cnidarian-algal symbiosis. Typically, each polyp harbors one species of algae. Via photosynthesis, these provide energy for the coral, and aid in calcification. As much as 30% of the tissue of a polyp may be plant material.

The algae benefit from a safe place to live and consume the polyp's carbon dioxide and nitrogenous waste. Due to the strain the algae can put on the polyp, stress on the coral often drives them to eject the algae. Mass ejections are known as coral bleaching, because the algae contribute to coral's brown coloration; other colors, however, are due to host coral pigments, such as green fluorescent proteins (GFPs). Ejection increases the polyp's chance of surviving short-term stress—they can regain algae, possibly of a different species at a later time. If the stressful conditions persist, the polyp eventually dies.

Reproduction

Corals can be both gonochoristic (unisexual) and hermaphroditic, each of which can reproduce sexually and asexually. Reproduction also allows coral to settle in new areas.

Sexual

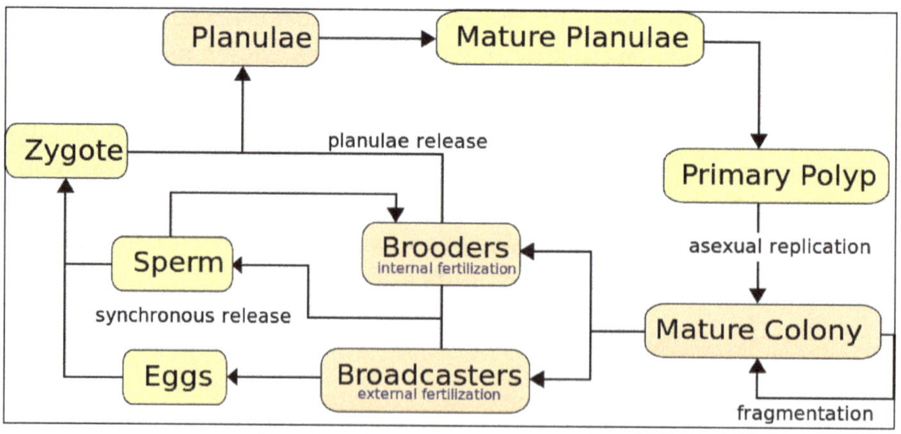

Life cycles of broadcasters and brooders.

Corals predominantly reproduce sexually. About 25% of hermatypic corals (stony corals) form single sex (gonochoristic) colonies, while the rest are hermaphroditic.

Broadcasters

About 75% of all hermatypic corals "broadcast spawn" by releasing gametes—eggs and sperm—into the water to spread offspring. The gametes fuse during fertilization to form a microscopic larva called a planula, typically pink and elliptical in shape. A typical coral colony forms several thousand larvae per year to overcome the odds against formation of a new colony.

A male great star coral, *Montastraea cavernosa*, releasing sperm into the water.

Synchronous spawning is very typical on the coral reef, and often, even when multiple species are present, all corals spawn on the same night. This synchrony is essential so male and female gametes can meet. Corals rely on environmental cues, varying from species to species, to determine the proper time to release gametes into the water. The cues involve temperature change, lunar cycle, day length, and possibly chemical signalling. Synchronous spawning may form hybrids and is perhaps involved in coral speciation. The immediate cue is most often sunset, which cues the release. The spawning event can be visually dramatic, clouding the usually clear water with gametes.

Brooders

Brooding species are most often ahermatypic (not reef-building) in areas of high current or wave action. Brooders release only sperm, which is negatively buoyant, sinking on to the waiting egg carriers who harbor unfertilized eggs for weeks. Synchronous spawning events sometimes occurs even with these species. After fertilization, the corals release planula that are ready to settle.

Planulae

Planulae exhibit positive phototaxis, swimming towards light to reach surface waters, where they drift and grow before descending to seek a hard surface to which they can attach and begin a new colony. They also exhibit positive sonotaxis, moving towards sounds that emanate from the reef and away from open water. High failure rates afflict many stages of this process, and even though millions of gametes are released by each colony, few new colonies form. The time from spawning to settling is usually two to three days, but can be up to two months. The larva grows into a polyp and eventually becomes a coral head by asexual budding and growth.

Asexual

Basal plates (calices) of *Orbicella annularis* showing multiplication by budding (small central plate) and division (large double plate).

Within a coral head, the genetically identical polyps reproduce asexually, either by budding (gemmation) or by dividing, whether longitudinally or transversely.

Budding involves splitting a smaller polyp from an adult. As the new polyp grows, it forms its body parts. The distance between the new and adult polyps grows, and with it, the coenosarc (the common body of the colony; coral anatomy). Budding can be intratentacular, from its oral discs, producing same-sized polyps within the ring of tentacles, or extratentacular, from its base, producing a smaller polyp.

Tabulate coral *Aulopora* (Devonian) showing initial budding.

Division forms two polyps that each become as large as the original. Longitudinal division begins when a polyp broadens and then divides its coelenteron (body), effectively splitting along its length. The mouth divides and new tentacles form. The two polyps thus created then generate their missing body parts and exoskeleton. Transversal division occurs when polyps and the exoskeleton divide transversally into two parts. This means one has the basal disc (bottom) and the other has the oral disc (top); the new polyps must separately generate the missing pieces.

Asexual reproduction offers the benefits of high reproductive rate, delaying senescence, and replacement of dead modules, as well as geographical distribution.

Colony Division

Whole colonies can reproduce asexually, forming two colonies with the same genotype. The possible mechanisms include fission, bailout and fragmentation. Fission occurs in some corals, especially among the family Fungiidae, where the colony splits into two or more colonies during early developmental stages. Bailout occurs when a single polyp abandons the colony and settles on a different substrate to create a new colony. Fragmentation involves individuals broken from the colony during storms or other disruptions. The separated individuals can start new colonies.

Reefs

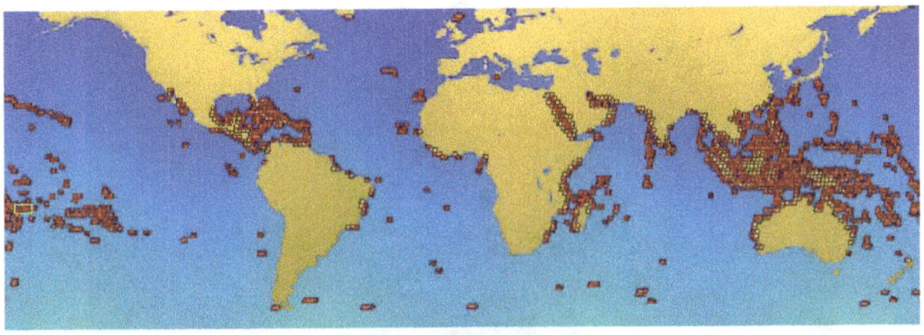

Locations of coral reefs around the world.

Many corals in the order Scleractinia are hermatypic, meaning that they are involved in building reefs. Most such corals obtain some of their energy from zooxanthellae in the genus *Symbiodinium*. These are symbiotic photosynthetic dinoflagellates which require sunlight; reef-forming corals are therefore found mainly in shallow water. They secrete calcium carbonate to form hard skeletons that become the framework of the reef. However, not all reef-building corals in shallow water contain zooxanthellae, and some deep water species, living at depths to which light cannot penetrate, form reefs but do not harbour the symbionts.

Staghorn coral (*Acropora cervicornis*) is an important hermatypic coral from the Caribbean.

There are various types of shallow-water coral reef, including fringing reefs, barrier reefs and atolls;

most occur in tropical and subtropical seas. They are very slow-growing, adding perhaps one centimetre (0.4 in) in height each year. The Great Barrier Reef is thought to have been laid down about two million years ago. Over time, corals fragment and die, sand and rubble accumulates between the corals, and the shells of clams and other molluscs decay to form a gradually evolving calcium carbonate structure. Coral reefs are extremely diverse marine ecosystems hosting over 4,000 species of fish, massive numbers of cnidarians, molluscs, crustaceans, and many other animals.

Evolutionary History

Solitary rugose coral (*Grewingkia*) in three views; Ordovician, southeastern Indiana.

Although corals first appeared in the Cambrian period, some 542 million years ago, fossils are extremely rare until the Ordovician period, 100 million years later, when rugose and tabulate corals became widespread. Paleozoic corals often contained numerous endobiotic symbionts.

Tabulate corals occur in limestones and calcareous shales of the Ordovician and Silurian periods, and often form low cushions or branching masses of calcite alongside rugose corals. Their numbers began to decline during the middle of the Silurian period, and they became extinct at the end of the Permian period, 250 million years ago.

Rugose or horn corals became dominant by the middle of the Silurian period, and became extinct early in the Triassic period. The rugose corals existed in solitary and colonial forms, and were also composed of calcite.

The scleractinian corals filled the niche vacated by the extinct rugose and tabulate species. Their fossils may be found in small numbers in rocks from the Triassic period, and became common in the Jurassic and later periods. Scleractinian skeletons are composed of a form of calcium carbonate known as aragonite. Although they are geologically younger than the tabulate and rugose corals, the aragonite of their skeletons is less readily preserved, and their fossil record is accordingly less complete.

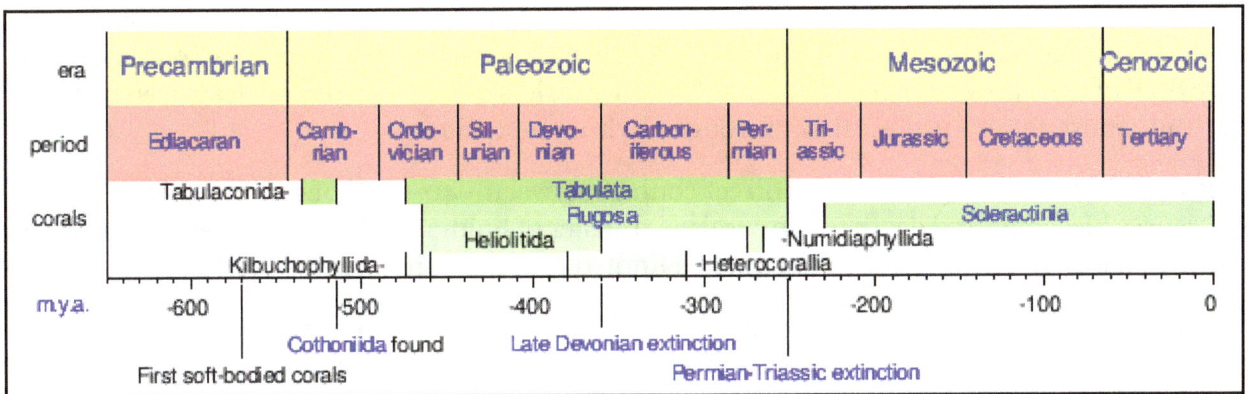

Timeline of the major coral fossil record and developments from 650 m.y.a. to present. edit

At certain times in the geological past, corals were very abundant. Like modern corals, these ancestors built reefs, some of which ended as great structures in sedimentary rocks. Fossils of fellow reef-dwellers algae, sponges, and the remains of many echinoids, brachiopods, bivalves, gastropods, and trilobites appear along with coral fossils. This makes some corals useful index fossils. Coral fossils are not restricted to reef remnants, and many solitary fossils may be found elsewhere, such as *Cyclocyathus*, which occurs in England's Gault clay formation.

Status

Threats

A healthy coral reef has a striking level of biodiversity in many forms of marine life.

Coral reefs are under stress around the world. In particular, coral mining, agricultural and urban runoff, pollution (organic and inorganic), overfishing, blast fishing, disease, and the digging of canals and access into islands and bays are localized threats to coral ecosystems. Broader threats are sea temperature rise, sea level rise and pH changes from ocean acidification, all associated with greenhouse gas emissions. In 1998, 16% of the world's reefs died as a result of increased water temperature.

Approximately 10% of the world's coral reefs are dead. About 60% of the world's reefs are at risk due to human-related activities. The threat to reef health is particularly strong in Southeast Asia, where 80% of reefs are endangered. Over 50% of the world's coral reefs may be destroyed by 2030; as a result, most nations protect them through environmental laws.

In the Caribbean and tropical Pacific, direct contact between ~40–70% of common seaweeds and coral causes bleaching and death to the coral via transfer of lipid-soluble metabolites. Seaweed and algae proliferate given adequate nutrients and limited grazing by herbivores such as parrotfish.

Water temperature changes of more than 1–2 °C (1.8–3.6 °F) or salinity changes can kill some species of coral. Under such environmental stresses, corals expel their Symbiodinium; without them coral tissues reveal the white of their skeletons, an event known as coral bleaching.

Submarine springs found along the coast of Mexico's Yucatán Peninsula produce water with a naturally low pH (relatively high acidity) providing conditions similar to those expected to become widespread as the oceans absorb carbon dioxide. Surveys discovered multiple species of live coral that appeared to tolerate the acidity. The colonies were small and patchily distributed, and had not formed structurally complex reefs such as those that compose the nearby Mesoamerican Barrier Reef System.

Protection

A diversity of corals.

Marine Protected Areas (MPAs), Biosphere reserves, marine parks, national monuments world heritage status, fishery management and habitat protection can protect reefs from anthropogenic damage.

Many governments now prohibit removal of coral from reefs, and inform coastal residents about reef protection and ecology. While local action such as habitat restoration and herbivore protection can reduce local damage, the longer-term threats of acidification, temperature change and sea-level rise remain a challenge.

To eliminate destruction of corals in their indigenous regions, projects have been started to grow corals in non-tropical countries.

Relation to Humans

Local economies near major coral reefs benefit from an abundance of fish and other marine creatures as a food source. Reefs also provide recreational scuba diving and snorkeling tourism. These activities can damage coral but international projects such as Green Fins that encourage dive and snorkel centres to follow a Code of Conduct have been proven to mitigate these risks.

Live coral is highly sought after for aquaria. Soft corals are easier to maintain in captivity than hard corals.

Jewelry

6-strand necklace, Navajo (Native American), ca. 1920s, Brooklyn Museum.

Corals' many colors give it appeal for necklaces and other jewelry. Intensely red coral is prized as a gemstone. Sometimes called fire coral, it is not the same as fire coral. Red coral is very rare because of overharvesting.

Medicine

In medicine, chemical compounds from corals are used for cancer, AIDS, pain, and other uses. Coral skeletons, e.g. *Isididae* are also used for bone grafting in humans. Coral Calx, known as Praval Bhasma in Sanskrit, is widely used in traditional system of Indian medicine as a supplement in the treatment of a variety of bone metabolic disorders associated with calcium deficiency.

Construction

Coral reefs in places such as the East African coast are used as a source of building material. Ancient (fossil) coral limestone, notably including the Coral Rag Formation of the hills around Oxford (England), was once used as a building stone, and can be seen in some of the oldest buildings in that city including the Saxon tower of St Michael at the Northgate, St. George's Tower of Oxford Castle, and the mediaeval walls of the city.

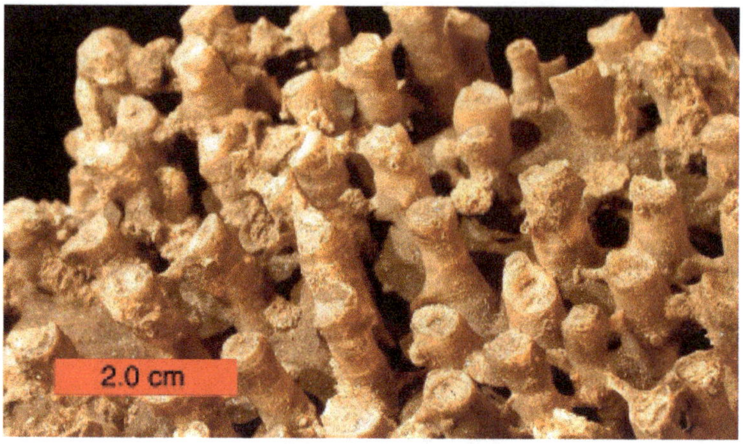

Tabulate coral (a syringoporid); Boone limestone (Lower Carboniferous) near Hiwasse, Arkansas, scale bar is 2.0 cm.

Climate Research

Annual growth bands in some corals, such as the deep sea bamboo corals (*Isididae*), may be among the first signs of the effects of ocean acidification on marine life. The growth rings allow geologists to construct year-by-year chronologies, a form of incremental dating, which underlie high-resolution records of past climatic and environmental changes using geochemical techniques.

Certain species form communities called microatolls, which are colonies whose top is dead and mostly above the water line, but whose perimeter is mostly submerged and alive. Average tide level limits their height. By analyzing the various growth morphologies, microatolls offer a low resolution record of sea level change. Fossilized microatolls can also be dated using Radiocarbon dating. Such methods can help to reconstruct Holocene sea levels.

Increasing sea temperatures in tropical regions (~1 degree C) the last century have caused major coral bleaching, death, and therefore shrinking coral populations since although they are able to adapt and acclimate, it is uncertain if this evolutionary process will happen quickly enough to prevent major reduction of their numbers.

Though coral have large sexually-reproducing populations, their evolution can be slowed by abundant asexual reproduction. Gene flow is variable among coral species. According to the biogeography of coral species gene flow cannot be counted on as a dependable source of adaptation as they are very stationary organisms. Also, coral longevity might factor into their adaptivity.

However, adaptation to climate changes has been demonstrated in many cases. These are usually due to a shift in coral and zooxanthellae genotypes. These shifts in allelic frequencies have progressed toward more tolerant types of zooxanthellae. Scientists found that a certain scleractinian zooxanthella is becoming more common where sea temperature is high. Symbionts able to tolerate warmer water seem to photosynthesise more slowly, implying an evolutionary trade-off.

In the Gulf of Mexico, where sea temperatures are rising, cold-sensitive staghorn and elkhorn coral have shifted in location. Not only have the symbionts and specific species been shown to shift, but there seems to be a certain growth rate favorable to selection. Slower-growing but more heat-tolerant corals have become more common. The changes in temperature and acclimation are complex. Some reefs in current shadows represent a refugium location that will help them adjust to the disparity in the environment even if eventually the temperatures may rise more quickly there than in other locations. This separation of populations by climatic barriers causes a realized niche to shrink greatly in comparison to the old fundamental niche.

Geochemistry

Corals are shallow, colonial organisms that integrate $\delta^{18}O$ and trace elements into their skeletal aragonite (polymorph of calcite) crystalline structures, as they grow. Geochemistry anomalies within the crystalline structures of corals represent functions of temperature, salinity and oxygen isotopic composition. Such geochemical analysis can help with climate modeling.

Strontium/Calcium Ratio Anomaly

Time can be attributed to coral geochemistry anomalies by correlating strontium/calcium minimums with sea surface temperature (SST) maximums to data collected from NINO 3.4 SSTA.

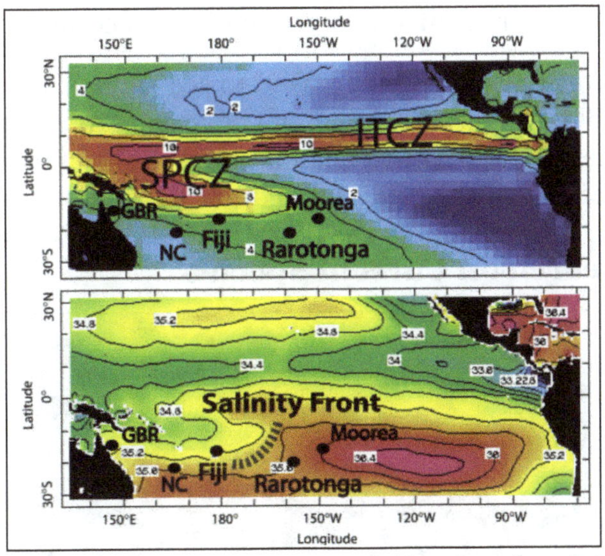

The position of the South Pacific convergence zone as a function of precipitation and salinity.

Oxygen Isotope Anomaly

The comparison of coral strontium/calcium minimums with sea surface temperature maximums, data recorded from NINO 3.4 SSTA, time can be correlated to coral strontium/calcium and $\delta^{18}O$ variations. To confirm accuracy of the annual relationship between Sr/Ca and $\delta^{18}O$ variations, a perceptible association to annual coral growth rings confirms the age conversion. Geochronology is established by the blending of Sr/Ca data, growth rings, and stable isotope data. El Nino-Southern Oscillation (ENSO) is directly related to climate fluctuations that influence coral $\delta^{18}O$ ratio from local salinity variations associated with the position of the South Pacific convergence zone (SPCZ) and can be used for ENSO modeling.

Sea Surface Temperature and Sea Surface Salinity

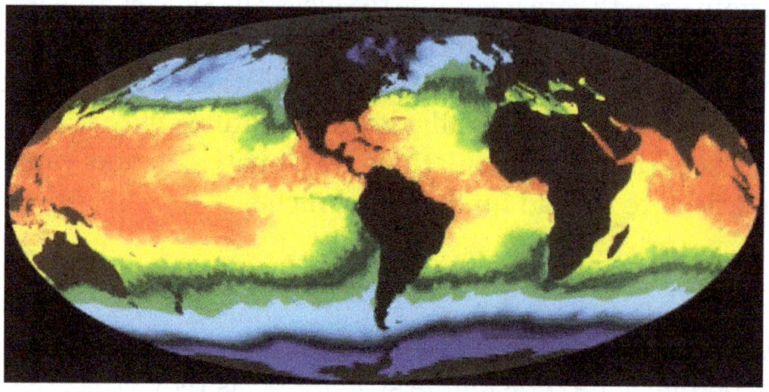

Global sea surface temperature (SST).

The global moisture budget is primarily being influenced by tropical sea surface temperatures from the position of the Intertropical Convergence Zone (ITCZ). The Southern Hemisphere has a unique meteorological feature positioned in the southwestern Pacific Basin called the South Pacific Convergence Zone (SPCZ), which contains a perennial position within the Southern Hemisphere. During ENSO warm periods, the SPCZ reverses orientation extending from the equator down south through Solomon Islands, Vanuatu, Fiji and towards the French Polynesian Islands; and due east towards South America affecting geochemistry of corals in tropical regions.

Geochemical analysis of skeletal coral can be linked to sea surface salinity (SSS) and sea surface temperature (SST), from El Nino 3.4 SSTA data, of tropical oceans to seawater $\delta^{18}O$ ratio anomalies from corals. ENSO phenomenon can be related to variations in sea surface salinity (SSS) and sea surface temperature (SST) that can help model tropical climate activities.

Limited Climate Research on Current Species

Genus: *Porites lutea*.

Climate research on live coral species is limited to a few studied species. Studying *Porites* coral provides a stable foundation for geochemical interpretations that is much simpler to physically extract data in comparison to *Platygyra* species where the complexity of *Platygyra* species skeletal structure creates difficulty when physically sampled, which happens to be one of the only multidecadal living coral records used for coral paleoclimate modeling.

Aquaria

This dragon-eye zoanthid is a popular source of color in reef tanks.

The saltwater fishkeeping hobby has increasingly expanded, over recent years, to include reef tanks, fish tanks that include large amounts of live rock on which coral is allowed to grow and spread. These tanks are either kept in a natural-like state, with algae (sometimes in the form of an algae scrubber) and a deep sand bed providing filtration, or as "show tanks", with the rock kept largely bare of the algae and microfauna that would normally populate it, in order to appear neat and clean.

The most popular kind of coral kept is soft coral, especially zoanthids and mushroom corals, which are especially easy to grow and propagate in a wide variety of conditions, because they originate in enclosed parts of reefs where water conditions vary and lighting may be less reliable and direct. More serious fishkeepers may keep small polyp stony coral, which is from open, brightly lit reef conditions and therefore much more demanding, while large polyp stony coral is a sort of compromise between the two.

Aquaculture

Coral aquaculture, also known as *coral farming* or *coral gardening*, is the cultivation of corals for commercial purposes or coral reef restoration. Aquaculture is showing promise as a potentially effective tool for restoring coral reefs, which have been declining around the world. The process bypasses the early growth stages of corals when they are most at risk of dying. Coral fragments known as "seeds" are grown in nurseries then replanted on the reef. Coral is farmed by coral farmers who live locally to the reefs and farm for reef conservation or for income. It is also farmed by scientists for research, by businesses for the supply of the live and ornamental coral trade and by private aquarium hobbyists.

Coral Reef

Coral reefs are diverse underwater ecosystems held together by calcium carbonate structures secreted by corals. Coral reefs are built by colonies of tiny animals found in marine waters that contain few nutrients. Most coral reefs are built from stony corals, which in turn consist of polyps that cluster in groups. The polyps belong to a group of animals known as Cnidaria, which also includes sea anemones and jellyfish. Unlike sea anemones, corals secrete hard carbonate exoskeletons which support and protect the coral polyps. Most reefs grow best in warm, shallow, clear, sunny and agitated waters.

Often called "rainforests of the sea", shallow coral reefs form some of the most diverse ecosystems on Earth. They occupy less than 0.1% of the world's ocean surface, about half the area of France, yet they provide a home for at least 25% of all marine species, including fish, mollusks, worms, crustaceans, echinoderms, sponges, tunicates and other cnidarians. Paradoxically, coral reefs flourish even though they are surrounded by ocean waters that provide few nutrients. They are most commonly found at shallow depths in tropical waters, but deep water and cold water corals also exist on smaller scales in other areas.

Coral reefs deliver ecosystem services to tourism, fisheries and shoreline protection. The annual global economic value of coral reefs is estimated between US$29.8-375 billion. However, coral reefs are fragile ecosystems, partly because they are very sensitive to water temperature. They are under threat from climate change, oceanic acidification, blast fishing, cyanide fishing for aquarium fish, sunscreen use, overuse of reef resources, and harmful land-use practices, including urban and agricultural runoff and water pollution, which can harm reefs by encouraging excess algal growth.

Formation

Most of the coral reefs we can see today were formed after the last glacial period when melting ice caused the sea level to rise and flood the continental shelves. This means that most modern coral reefs are less than 10,000 years old. As communities established themselves on the shelves, the reefs grew upwards, pacing rising sea levels. Reefs that rose too slowly could become drowned reefs. They are covered by so much water that there was insufficient light. Coral reefs are found in the deep sea away from continental shelves, around oceanic islands and as atolls. The vast majority of these islands are volcanic in origin. The few exceptions have tectonic origins where plate movements have lifted the deep ocean floor on the surface.

In 1842 in his first monograph, *The Structure and Distribution of Coral Reefs*, Charles Darwin set out his theory of the formation of atoll reefs, an idea he conceived during the voyage of the *Beagle*. He theorized uplift and subsidence of the Earth's crust under the oceans formed the atolls. Darwin's theory sets out a sequence of three stages in atoll formation. It starts with a fringing reef forming around an extinct volcanic island as the island and ocean floor subsides. As the subsidence continues, the fringing reef becomes a barrier reef, and ultimately an atoll reef.

Darwin's theory starts with a volcanic island which becomes extinct.

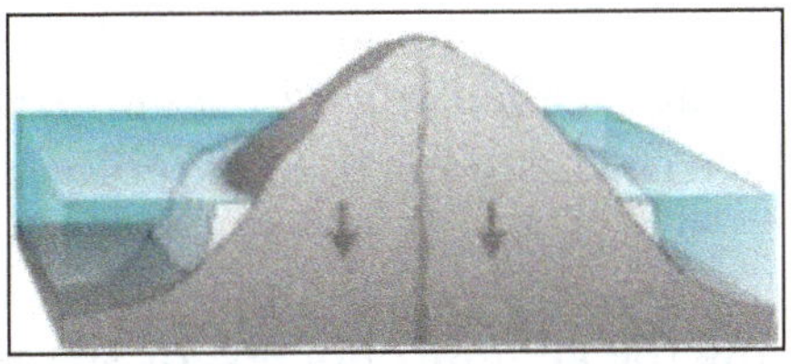

As the island and ocean floor subside, coral growth builds a fringing reef, often including a shallow lagoon between the land and the main reef.

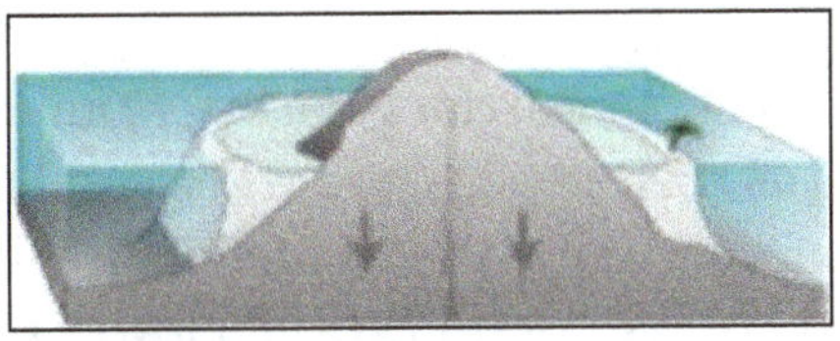

As the subsidence continues, the fringing reef becomes a larger barrier reef further from the shore with a bigger and deeper lagoon inside.

Ultimately, the island sinks below the sea, and the barrier reef becomes an atoll enclosing an open lagoon.

Darwin predicted that underneath each lagoon would be a bed rock base, the remains of the original volcano. Subsequent drilling proved this correct. Darwin's theory followed from his understanding that coral polyps thrive in the clean seas of the tropics where the water is agitated, but can only live within a limited depth range, starting just below low tide. Where the level of the underlying earth allows, the corals grow around the coast to form what he called fringing reefs, and can eventually grow out from the shore to become a barrier reef.

A fringing reef can take ten thousand years to form, and an atoll can take up to 30 million years.

Where the bottom is rising, fringing reefs can grow around the coast, but coral raised above sea level dies and becomes white limestone. If the land subsides slowly, the fringing reefs keep pace by growing upwards on a base of older, dead coral, forming a barrier reef enclosing a lagoon between the reef and the land. A barrier reef can encircle an island, and once the island sinks below sea level a roughly circular atoll of growing coral continues to keep up with the sea level, forming a central lagoon. Barrier reefs and atolls do not usually form complete circles, but are broken in places by storms. Like sea level rise, a rapidly subsiding bottom can overwhelm coral growth, killing the coral polyps and the reef, due to what is called *coral drowning*. Corals that rely on zooxanthellae can *drown* when the water becomes too deep for their symbionts to adequately photosynthesize, due to decreased light exposure.

The two main variables determining the geomorphology, or shape, of coral reefs are the nature of the underlying substrate on which they rest, and the history of the change in sea level relative to that substrate.

The approximately 20,000-year-old Great Barrier Reef offers an example of how coral reefs formed on continental shelves. Sea level was then 120 m (390 ft) lower than in the 21st century. As sea level rose, the water and the corals encroached on what had been hills of the Australian coastal plain. By 13,000 years ago, sea level had risen to 60 m (200 ft) lower than at present, and many hills of the coastal plains had become continental islands. As the sea level rise continued, water topped most of the continental islands. The corals could then overgrow the hills, forming the present cays and reefs. Sea level on the Great Barrier Reef has not changed significantly in the last 6,000 years, and the age of the modern living reef structure is estimated to be between 6,000 and 8,000 years. Although the Great Barrier Reef formed along a continental shelf, and not around a volcanic island, Darwin's principles apply. Development stopped at the barrier reef stage, since Australia is not about to submerge. It formed the world's largest barrier reef, 300–1,000 m (980–3,280 ft) from shore, stretching for 2,000 km (1,200 mi).

Healthy tropical coral reefs grow horizontally from 1 to 3 cm (0.39 to 1.18 in) per year, and grow vertically anywhere from 1 to 25 cm (0.39 to 9.84 in) per year; however, they grow only at depths shallower than 150 m (490 ft) because of their need for sunlight, and cannot grow above sea level.

Materials

As the name implies, the bulk of coral reefs is made up of coral skeletons from mostly intact coral colonies. As other chemical elements present in corals become incorporated into the calcium carbonate deposits, aragonite is formed. However, shell fragments and the remains of calcareous algae such as the green-segmented genus *Halimeda* can add to the reef's ability to withstand damage from storms and other threats. Such mixtures are visible in structures such as Eniwetok Atoll.

Types

The three principal reef types are:

- Fringing reef – directly attached to a shore, or borders it with an intervening shallow channel or lagoon.
- Barrier reef – reef separated from a mainland or island shore by a deep channel or lagoon.
- Atoll reef – more or less circular or continuous barrier reef extends all the way around a lagoon without a central island.

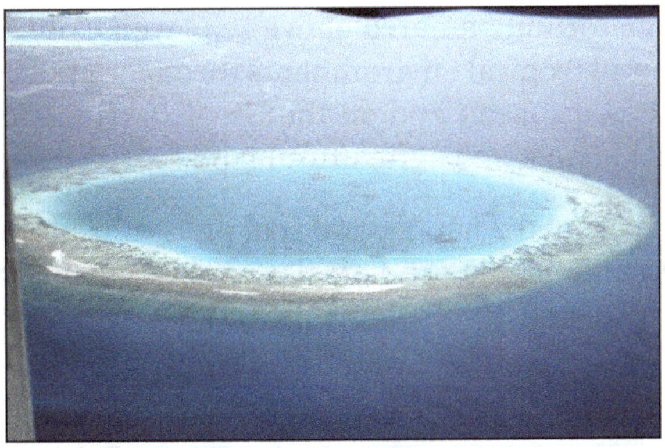

A small atoll in the Maldives.

Inhabited cay in the Maldives.

Other reef types or variants are:

- Patch reef – common, isolated, comparatively small reef outcrop, usually within a lagoon or embayment, often circular and surrounded by sand or seagrass.
- Apron reef – short reef resembling a fringing reef, but more sloped; extending out and downward from a point or peninsular shore.
- Bank reef – linear or semicircular shaped-outline, larger than a patch reef.
- Ribbon reef – long, narrow, possibly winding reef, usually associated with an atoll lagoon.
- Table reef – isolated reef, approaching an atoll type, but without a lagoon.
- Habili – reef specific to the Red Sea; does not reach the surface near enough to cause visible surf; may be a hazard to ships (from the Arabic for "unborn").
- Microatoll – community of species of corals; vertical growth limited by average tidal height; growth morphologies offer a low-resolution record of patterns of sea level change; fossilized remains can be dated using radioactive carbon dating and have been used to reconstruct Holocene sea levels.
- Cays – small, low-elevation, sandy islands formed on the surface of coral reefs from eroded

material that piles up, forming an area above sea level; can be stabilized by plants to become habitable; occur in tropical environments throughout the Pacific, Atlantic and Indian Oceans (including the Caribbean and on the Great Barrier Reef and Belize Barrier Reef), where they provide habitable and agricultural land.

- Seamount or guyot – formed when a coral reef on a volcanic island subsides; tops of seamounts are rounded and guyots are flat; flat tops of guyots, or *tablemounts*, are due to erosion by waves, winds, and atmospheric processes.

Zones

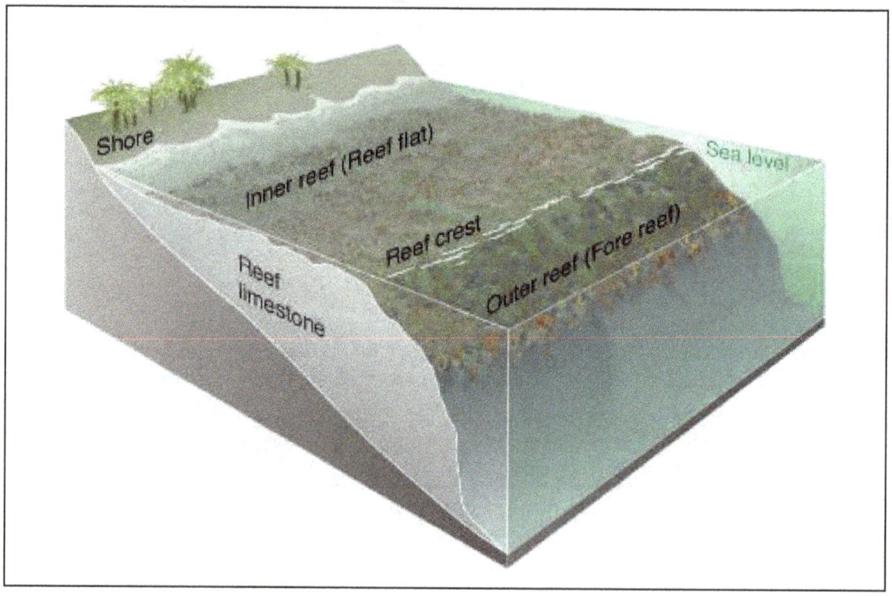

The three major zones of a coral reef: the fore reef, reef crest, and the back reef.

Coral reef ecosystems contain distinct zones that represent different kinds of habitats. Usually, three major zones are recognized: the fore reef, reef crest, and the back reef (frequently referred to as the reef lagoon).

All three zones are physically and ecologically interconnected. Reef life and oceanic processes create opportunities for exchange of seawater, sediments, nutrients, and marine life among one another.

Thus, they are integrated components of the coral reef ecosystem, each playing a role in the support of the reefs' abundant and diverse fish assemblages.

Most coral reefs exist in shallow waters less than 50 m deep. Some inhabit tropical continental shelves where cool, nutrient rich upwelling does not occur, such as Great Barrier Reef. Others are found in the deep ocean surrounding islands or as atolls, such as in the Maldives. The reefs surrounding islands form when islands subside into the ocean, and atolls form when an island subsides below the surface of the sea.

Alternatively, Moyle and Cech distinguish six zones, though most reefs possess only some of the zones.

Water in the reef surface zone is often agitated. This diagram represents a reef on a continental shelf. The water waves at the left travel over the *off-reef floor* until they encounter the *reef slope* or *fore reef*. Then the waves pass over the shallow *reef crest*. When a wave enters shallow water it shoals, that is, it slows down and the wave height increases.

The reef surface is the shallowest part of the reef. It is subject to the surge and the rise and fall of tides. When waves pass over shallow areas, they shoal, as shown in the diagram at the right. This means the water is often agitated. These are the precise condition under which corals flourish. Shallowness means there is plenty of light for photosynthesis by the symbiotic zooxanthellae, and agitated water promotes the ability of coral to feed on plankton. However, other organisms must be able to withstand the robust conditions to flourish in this zone.

The off-reef floor is the shallow sea floor surrounding a reef. This zone occurs by reefs on continental shelves. Reefs around tropical islands and atolls drop abruptly to great depths, and do not have a floor. Usually sandy, the floor often supports seagrass meadows which are important foraging areas for reef fish.

The reef drop-off is, for its first 50 m, habitat for many reef fish who find shelter on the cliff face and plankton in the water nearby. The drop-off zone applies mainly to the reefs surrounding oceanic islands and atolls.

The reef face is the zone above the reef floor or the reef drop-off. This zone is often the most diverse area of the reef. Coral and calcareous algae growths provide complex habitats and areas which offer protection, such as cracks and crevices. Invertebrates and epiphytic algae provide much of the food for other organisms.

The reef flat is the sandy-bottomed flat, which can be behind the main reef, containing chunks of coral. This zone may border a lagoon and serve as a protective area, or it may lie between the reef and the shore, and in this case is a flat, rocky area. Fishes tend to prefer living in that flat, rocky area, compared to any other zone, when it is present.

The reef lagoon is an entirely enclosed region, which creates an area less affected by wave action that often contains small reef patches.

However, the "topography of coral reefs is constantly changing. Each reef is made up of irregular patches of algae, sessile invertebrates, and bare rock and sand. The size, shape and relative abundance of these patches changes from year to year in response to the various factors that favor one type of patch over another. Growing coral, for example, produces constant change in the fine structure of reefs. On a larger scale, tropical storms may knock out large sections of reef and cause boulders on sandy areas to move."

Locations

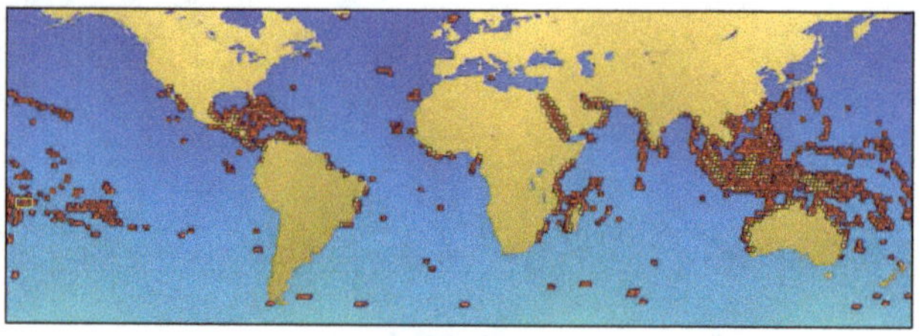

Locations of coral reefs.

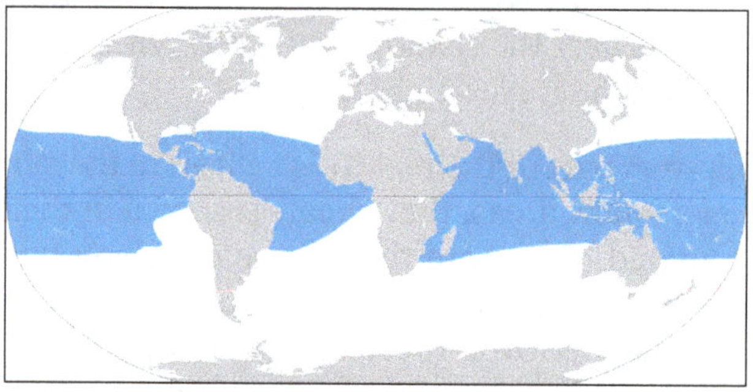

Boundary for 20 °C isotherms. Most corals live within this boundary. Note the cooler waters caused by upwelling on the southwest coast of Africa and off the coast of Peru.

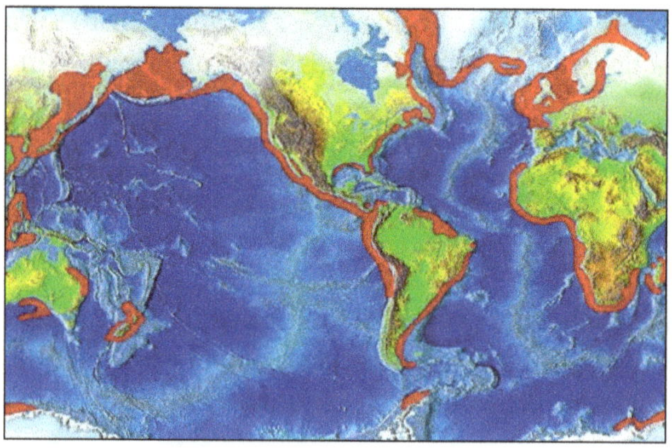

This map shows areas of upwelling in red. Coral reefs are not found in coastal areas where colder and nutrient-rich upwellings occur.

Coral reefs are estimated to cover 284,300 km^2 (109,800 sq mi), just under 0.1% of the oceans' surface area. The Indo-Pacific region (including the Red Sea, Indian Ocean, Southeast Asia and the Pacific) account for 91.9% of this total. Southeast Asia accounts for 32.3% of that figure, while the Pacific including Australia accounts for 40.8%. Atlantic and Caribbean coral reefs account for 7.6%.

Although corals exist both in temperate and tropical waters, shallow-water reefs form only in a zone extending from approximately 30° N to 30° S of the equator. Tropical corals do not grow at depths of over 50 meters (160 ft). The optimum temperature for most coral reefs is 26–27 °C

(79–81 °F), and few reefs exist in waters below 18 °C (64 °F). However, reefs in the Persian Gulf have adapted to temperatures of 13 °C (55 °F) in winter and 38 °C (100 °F) in summer. There are 37 species of scleractinian corals identified in such harsh environment around Larak Island.

Deep-water coral can exist at greater depths and colder temperatures at much higher latitudes, as far north as Norway. Although deep water corals can form reefs, very little is known about them.

Coral reefs are rare along the west coasts of the Americas and Africa, due primarily to upwelling and strong cold coastal currents that reduce water temperatures in these areas (respectively the Peru, Benguela and Canary streams). Corals are seldom found along the coastline of South Asia—from the eastern tip of India (Chennai) to the Bangladesh and Myanmar borders—as well as along the coasts of northeastern South America and Bangladesh, due to the freshwater release from the Amazon and Ganges Rivers respectively.

- The Great Barrier Reef—largest, comprising over 2,900 individual reefs and 900 islands stretching for over 2,600 kilometers (1,600 mi) off Queensland, Australia.
- The Mesoamerican Barrier Reef System—second largest, stretching 1,000 kilometers (620 mi) from Isla Contoy at the tip of the Yucatán Peninsula down to the Bay Islands of Honduras.
- The New Caledonia Barrier Reef—second longest double barrier reef, covering 1,500 kilometers (930 mi).
- The Andros, Bahamas Barrier Reef—third largest, following the east coast of Andros Island, Bahamas, between Andros and Nassau.
- The Red Sea—includes 6000-year-old fringing reefs located around a 2,000 km (1,240 mi) coastline.
- The Florida Reef Tract—largest continental US reef, extends from Soldier Key, located in Biscayne Bay, to the Dry Tortugas in the Gulf of Mexico.
- Pulley Ridge—deepest photosynthetic coral reef, Florida.
- Numerous reefs scattered over the Maldives.
- The Philippines coral reef area, the second largest in Southeast Asia, is estimated at 26,000 square kilometers and holds an extraordinary diversity of species. Scientists have identified 915 reef fish species and more than 400 scleractinian coral species, 12 of which are endemic.
- The Raja Ampat Islands in Indonesia's West Papua province offer the highest known marine diversity.
- Bermuda is known for its northernmost coral reef system, located at 32.4° N and 64.8° W. The presence of coral reefs at this high latitude is due to the proximity of the Gulf Stream. Bermuda has a fairly consistent diversity of coral species, representing a subset of those found in the greater Caribbean.
- The world's northernmost individual coral reef so far discovered is located within a bay of Japan's Tsushima Island in the Korea Strait.

- The world's southernmost coral reef is at Lord Howe Island, in the Pacific Ocean off the east coast of Australia.

Biology

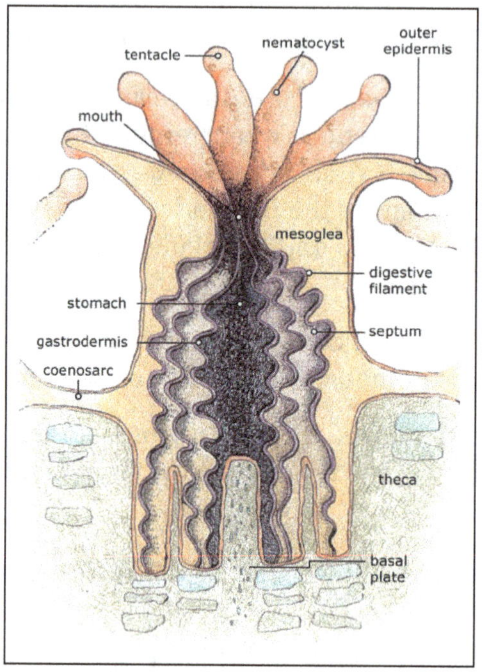

Anatomy of a coral polyp.

Alive corals are colonies of small animals embedded in calcium carbonate shells. It is a mistake to think of coral as plants or rocks. Coral heads consist of accumulations of individual animals called polyps, arranged in diverse shapes. Polyps are usually tiny, but they can range in size from a pin-head to 12 inches (30 cm) across.

Reef-building or hermatypic corals live only in the photic zone (above 50 m), the depth to which sufficient sunlight penetrates the water, allowing photosynthesis to occur. Coral polyps do not photosynthesize, but have a symbiotic relationship with microscopic algae of the genus *Symbiodinium*, commonly referred to as zooxanthellae. These organisms live within the tissues of polyps and provide organic nutrients that nourish the polyp. Because of this relationship, coral reefs grow much faster in clear water, which admits more sunlight. Without their symbionts, coral growth would be too slow to form significant reef structures. Corals get up to 90% of their nutrients from their symbionts.

Reefs grow as polyps and other organisms deposit calcium carbonate, the basis of coral, as a skeletal structure beneath and around themselves, pushing the coral head's top upwards and outwards. Waves, grazing fish (such as parrotfish), sea urchins, sponges, and other forces and organisms act as bioeroders, breaking down coral skeletons into fragments that settle into spaces in the reef structure or form sandy bottoms in associated reef lagoons. Many other organisms living in the reef community contribute skeletal calcium carbonate in the same manner. Coralline algae are important contributors to reef structure in those parts of the reef subjected to the greatest forces by waves (such as the reef front facing the open ocean). These algae strengthen the reef structure by depositing limestone in sheets over the reef surface.

Typical shapes for coral species are wrinkled brains, cabbages, table tops, antlers, wire strands and pillars. These shapes can depend on the life history of the coral, like light exposure and wave action, and events such as breakages.

Table coral.

Corals reproduce both sexually and asexually. An individual polyp uses both reproductive modes within its lifetime. Corals reproduce sexually by either internal or external fertilization. The reproductive cells are found on the mesenteries, membranes that radiate inward from the layer of tissue that lines the stomach cavity. Some mature adult corals are hermaphroditic; others are exclusively male or female. A few species change sex as they grow.

Close up of polyps are arrayed on a coral, waving their tentacles. There can be thousands of polyps on a single coral branch.

Internally fertilized eggs develop in the polyp for a period ranging from days to weeks. Subsequent development produces a tiny larva, known as a planula. Externally fertilized eggs develop during synchronized spawning. Polyps release eggs and sperm into the water en masse, simultaneously. Eggs disperse over a large area. The timing of spawning depends on time of year, water temperature, and tidal and lunar cycles. Spawning is most successful when there is little variation between high and low tide. The less water movement, the better the chance for fertilization. Ideal timing occurs in the spring. Release of eggs or planula usually occurs at night, and is sometimes in phase with the lunar cycle (three to six days after a full moon). The period from release to settlement lasts only a few days, but some planulae can survive afloat for several weeks. They are vulnerable to predation and environmental conditions. The lucky few planulae which successfully attach to substrate next confront competition for food and space.

There are eight clades of Symbiodinium phylotypes. Most research has been completed on the Symbiodinium clades A–D. Each one of the eight contributes their own benefits as well as less compatible attributes to the survival of their coral hosts. Each photosynthetic organism has a specific level of sensitivity to photodamage of compounds needed for survival, such as proteins. Rates of regeneration and replication determine the organism's ability to survive. Phylotype A is found more in the shallow regions of marine waters. It is able to produce mycosporine-like amino acids that are UV resistant, using a derivative of glycerin to absorb the UV radiation and allowing them to become more receptive to warmer water temperatures. In the event of UV or thermal damage, if and when repair occurs, it will increase the likelihood of survival of the host and symbiont. This leads to the idea that, evolutionarily, clade A is more UV resistant and thermally resistant than the other clades.

Clades B and C are found more frequently in the deeper water regions, which may explain the higher susceptibility to increased temperatures. Terrestrial plants that receive less sunlight because they are found in the undergrowth can be analogized to clades B, C, and D. Since clades B through D are found at deeper depths, they require an elevated light absorption rate to be able to synthesize as much energy. With elevated absorption rates at UV wavelengths, the deeper occurring phylotypes are more prone to coral bleaching versus the more shallow clades. Clade D has been observed to be high temperature-tolerant, and as a result it has a higher rate of survival than clades B and C.

Black coral.

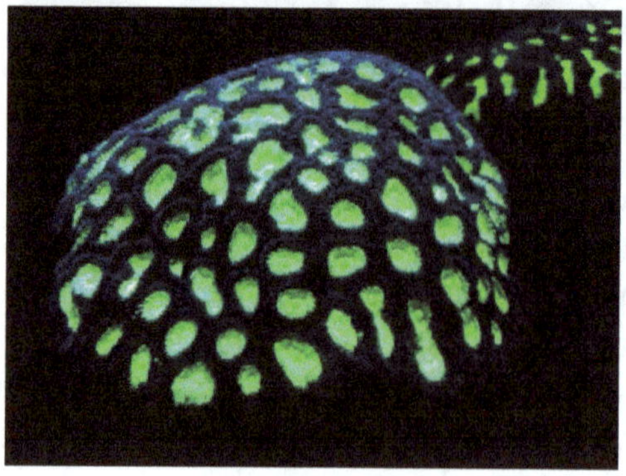

Fluorescent coral.

Darwin's Paradox

"Coral... seems to proliferate when ocean waters are warm, poor, clear and agitated, a fact which Darwin had already noted when he passed through Tahiti in 1842. This constitutes a fundamental paradox, shown quantitatively by the apparent impossibility of balancing input and output of the nutritive elements which control the coral polyp metabolism.

Recent oceanographic research has brought to light the reality of this paradox by confirming that the oligotrophy of the ocean euphotic zone persists right up to the swell-battered reef crest. When you approach the reef edges and atolls from the quasidesert of the open sea, the near absence of living matter suddenly becomes a plethora of life, without transition. So why is there something rather than nothing, and more precisely, where do the necessary nutrients for the functioning of this extraordinary coral reef machine come from?" — Francis Rougerie.

In *The Structure and Distribution of Coral Reefs*, published in 1842, Darwin described how coral reefs were found in some areas of the tropical seas but not others, with no obvious cause. The largest and strongest corals grew in parts of the reef exposed to the most violent surf and corals were weakened or absent where loose sediment accumulated.

Tropical waters contain few nutrients yet a coral reef can flourish like an "oasis in the desert". This has given rise to the ecosystem conundrum, sometimes called "Darwin's paradox": "How can such high production flourish in such nutrient poor conditions?"

Coral reefs cover less than 0.1% of the surface of the world's ocean, about half the land area of France, yet they support over one-quarter of all marine species. This diversity results in complex food webs, with large predator fish eating smaller forage fish that eat yet smaller zooplankton and so on. However, all food webs eventually depend on plants, which are the primary producers. Coral reefs' primary productivity is very high, typically producing 5–10 grams of carbon per square meter per day ($gC \cdot m^{-2} \cdot day^{-1}$) biomass.

Coral Polyps.

One reason for the unusual clarity of tropical waters is they are deficient in nutrients and drifting plankton. Further, the sun shines year round in the tropics, warming the surface layer, making it less dense than subsurface layers. The warmer water is separated from deeper, cooler water by a stable thermocline, where the temperature makes a rapid change. This keeps the warm surface waters floating above the cooler deeper waters. In most parts of the ocean, there is little exchange between these layers. Organisms that die in aquatic environments generally sink to the bottom, where they decompose, which releases nutrients in the form of nitrogen (N), phosphorus (P) and potassium (K). These nutrients are necessary for plant growth, but in the tropics, they do not directly return to the surface.

Plants form the base of the food chain, and need sunlight and nutrients to grow. In the ocean, these plants are mainly microscopic phytoplankton which drift in the water column. They need sunlight for photosynthesis, which powers carbon fixation, so they are found only relatively near the surface. But they also need nutrients. Phytoplankton rapidly use nutrients in the surface waters, and in the tropics, these nutrients are not usually replaced because of the thermocline.

Explanations

Around coral reefs, lagoons fill in with material eroded from the reef and the island. They become havens for marine life, providing protection from waves and storms.

Most importantly, reefs recycle nutrients, which happens much less in the open ocean. In coral reefs and lagoons, producers include phytoplankton, as well as seaweed and coralline algae, especially small types called turf algae, which pass nutrients to corals. The phytoplankton are eaten by fish and crustaceans, who also pass nutrients along the food web. Recycling ensures fewer nutrients are needed overall to support the community.

Coral reefs support many symbiotic relationships. In particular, zooxanthellae provide energy to coral in the form of glucose, glycerol, and amino acids. Zooxanthellae can provide up to 90% of a coral's energy requirements. In return, as an example of mutualism, the corals shelter the zooxanthellae, averaging one million for every cubic centimeter of coral, and provide a constant supply of the carbon dioxide they need for photosynthesis.

The color of corals depends on the combination of brown shades provided by their zooxanthellae and pigmented proteins (reds, blues, greens, etc.) produced by the corals themselves.

Corals also absorb nutrients, including inorganic nitrogen and phosphorus, directly from water. Many corals extend their tentacles at night to catch zooplankton that brush them when the water is agitated. Zooplankton provide the polyp with nitrogen, and the polyp shares some of the nitrogen with the zooxanthellae, which also require this element. The varying pigments in different species of zooxanthellae give them an overall brown or golden-brown appearance, and give brown corals their colors. Other pigments such as reds, blues, greens, etc. come from colored proteins made by the coral animals. Coral which loses a large fraction of its zooxanthellae becomes white (or sometimes pastel shades in corals that are richly pigmented with their own colorful proteins) and is said to be bleached, a condition which, unless corrected, can kill the coral.

Sponges are another key: they live in crevices in the coral reefs. They are efficient filter feeders, and in the Red Sea they consume about 60% of the phytoplankton that drifts by. The sponges eventually excrete nutrients in a form the corals can use.

Most coral polyps are nocturnal feeders. Here, in the dark, polyps have extended their tentacles to feed on zooplankton.

The roughness of coral surfaces is the key to coral survival in agitated waters. Normally, a boundary layer of still water surrounds a submerged object, which acts as a barrier. Waves breaking on the extremely rough edges of corals disrupt the boundary layer, allowing the corals access to passing nutrients. Turbulent water thereby promotes reef growth and branching. Without the nutritional gains brought by rough coral surfaces, even the most effective recycling would leave corals wanting in nutrients.

Studies have shown that deep nutrient-rich water entering coral reefs through isolated events may have significant effects on temperature and nutrient systems. This water movement disrupts the relatively stable thermocline that usually exists between warm shallow water to deeper colder water. Leichter et al. (2006) found that temperature regimes on coral reefs in the Bahamas and Florida were highly variable with temporal scales of minutes to seasons and spatial scales across depths.

Water can be moved through coral reefs in various ways, including current rings, surface waves, internal waves and tidal changes. Movement is generally created by tides and wind. As tides interact with varying bathymetry and wind mixes with surface water, internal waves are created. An internal wave is a gravity wave that moves along density stratification within the ocean. When a water parcel encounters a different density it will oscillate and create internal waves. While internal waves generally have a lower frequency than surface waves, they often form as a single wave that breaks into multiple waves as it hits a slope and moves upward. This vertical break up of internal waves causes significant diapycnal mixing and turbulence. Internal waves can act as nutrient pumps, bringing plankton and cool nutrient-rich water up to the surface.

The irregular structure characteristic of coral reef bathymetry may enhance mixing and produce pockets of cooler water and variable nutrient content. Arrival of cool, nutrient-rich water from depths due to internal waves and tidal bores has been linked to growth rates of suspension feeders and benthic algae as well as plankton and larval organisms. Leichter et al. proposed that Codium isthmocladum react to deep water nutrient sources due to their tissues having different concentrations of nutrients dependent upon depth. Wolanski and Hamner noted aggregations of eggs, larval organisms and plankton on reefs in response to deep water intrusions. Similarly, as internal waves and bores move vertically, surface-dwelling larval organisms are carried toward the shore. This has significant biological importance to cascading effects of food chains in coral reef ecosystems and may provide yet another key to unlocking "Darwin's Paradox".

Cyanobacteria provide soluble nitrates for the reef via nitrogen fixation.

Coral reefs also often depend on surrounding habitats, such as seagrass meadows and mangrove forests, for nutrients. Seagrass and mangroves supply dead plants and animals which are rich in nitrogen and also serve to feed fish and animals from the reef by supplying wood and vegetation. Reefs, in turn, protect mangroves and seagrass from waves and produce sediment in which the mangroves and seagrass can root.

Biodiversity

Tube sponges attracting cardinal fishes, glassfishes and wrasses.

Coral reefs form some of the world's most productive ecosystems, providing complex and varied marine habitats that support a wide range of other organisms.[Fringing reefs] just below low tide level have a mutually beneficial relationship with mangrove forests at high tide level and sea grass meadows in between: the reefs protect the mangroves and seagrass from strong currents and waves that would damage them or erode the sediments in which they are rooted, while the mangroves and sea grass protect the coral from large influxes of silt, fresh water and pollutants. This level of variety in the environment benefits many coral reef animals, which, for example, may feed in the sea grass and use the reefs for protection or breeding.

Reefs are home to a large variety of animals, including fish, seabirds, sponges, cnidarians (which includes some types of corals and jellyfish), worms, crustaceans (including shrimp, cleaner shrimp, spiny lobsters and crabs), mollusks (including cephalopods), echinoderms (including starfish, sea urchins and sea cucumbers), sea squirts, sea turtles and sea snakes. Aside from humans, mammals are rare on coral reefs, with visiting cetaceans such as dolphins being the main exception. A few of these varied species feed directly on corals, while others graze on algae on the reef. Reef biomass is positively related to species diversity.

Organisms can cover every square inch of a coral reef.

The same hideouts in a reef may be regularly inhabited by different species at different times of day. Nighttime predators such as cardinalfish and squirrelfish hide during the day, while damselfish, surgeonfish, triggerfish, wrasses and parrotfish hide from eels and sharks.

Algae

Reefs are chronically at risk of algal encroachment. Overfishing and excess nutrient supply from onshore can enable algae to outcompete and kill the coral. Increased nutrient levels can be a result of sewage or chemical fertilizer runoff from nearby coastal developments. Runoff can carry nitrogen and phosphorus which promote excess algae growth. Algae can sometimes out-compete the coral for space. The algae can then smother the coral by decreasing the oxygen supply available to the reef. Decreased oxygen levels can slow down coral's calcification rates weakening the coral and leaving it more susceptible to disease and degradation. In surveys done around largely uninhabited US Pacific islands, algae inhabit a large percentage of surveyed coral locations. The algal population consists of turf algae, coralline algae, and macro algae.

Sponges

Sponges are essential for the functioning of the coral reef's ecosystem. Algae and corals in coral reefs produce organic material. This is filtered through sponges which convert this organic material into small particles which in turn are absorbed by algae and corals.

Fish

Over 4,000 species of fish inhabit coral reefs. The reasons for this diversity remain unclear. Hypotheses include the "lottery", in which the first (lucky winner) recruit to a territory is typically able to defend it against latecomers, "competition", in which adults compete for territory, and less-competitive species must be able to survive in poorer habitat, and "predation", in which population size is a function of postsettlement piscivore mortality. Healthy reefs can produce up to 35 tons of fish per square kilometer each year, but damaged reefs produce much less.

Invertebrates

Sea urchins, Dotidae and sea slugs eat seaweed. Some species of sea urchins, such as *Diadema antillarum*, can play a pivotal part in preventing algae from overrunning reefs. Nudibranchia and sea anemones eat sponges.

A number of invertebrates, collectively called "cryptofauna," inhabit the coral skeletal substrate itself, either boring into the skeletons (through the process of bioerosion) or living in pre-existing voids and crevices. Those animals boring into the rock include sponges, bivalve mollusks, and sipunculans. Those settling on the reef include many other species, particularly crustaceans and polychaete worms.

Seabirds

Coral reef systems provide important habitats for seabird species, some endangered. For example, Midway Atoll in Hawaii supports nearly three million seabirds, including two-thirds (1.5 million) of the global population of Laysan albatross, and one-third of the global population of black-footed albatross. Each seabird species has specific sites on the atoll where they nest. Altogether, 17 species of seabirds live on Midway. The short-tailed albatross is the rarest, with fewer than 2,200 surviving after excessive feather hunting in the late 19th century.

Other

Sea snakes feed exclusively on fish and their eggs. Marine birds, such as herons, gannets, pelicans and boobies, feed on reef fish. Some land-based reptiles intermittently associate with reefs, such as monitor lizards, the marine crocodile and semiaquatic snakes, such as *Laticauda colubrina*. Sea turtles, particularly hawksbill sea turtles, feed on sponges.

Schooling reef fish.

Importance

Coral reefs deliver ecosystem services to tourism, fisheries and coastline protection. The global economic value of coral reefs has been estimated to be between US $29.8 billion and $375 billion per year. Coral reefs protect shorelines by absorbing wave energy, and many small islands would not exist without their reefs to protect them. According to the environmental group World Wide Fund for Nature, the economic cost over a 25-year period of destroying one kilometer of coral reef is somewhere between $137,000 and $1,200,000. About six million tons of fish are taken each year from coral reefs. Well-managed coral reefs have an annual yield of 15 tons of seafood on average per square kilometer. Southeast Asia's coral reef fisheries alone yield about $2.4 billion annually from seafood.

To improve the management of coastal coral reefs, another environmental group, the World Resources Institute (WRI) developed and published tools for calculating the value of coral reef-related tourism, shoreline protection and fisheries, partnering with five Caribbean countries. As of April 2011, published working papers covered St. Lucia, Tobago, Belize, and the Dominican Republic, with a paper for Jamaica in preparation. The WRI was also "making sure that the study results support improved coastal policies and management planning". The Belize study estimated the value of reef and mangrove services at $395–559 million annually.

Bermuda's coral reefs provide economic benefits to the Island worth on average $722 million per year, based on six key ecosystem services, according to Sarkis *et al* (2010).

Threats

Island with fringing reef off Yap, Micronesia.

Coral reefs are dying around the world. In particular, coral mining, agricultural and urban runoff, pollution (organic and inorganic), overfishing, blast fishing, disease, and the digging of canals and access into islands and bays are localized threats to coral ecosystems. Broader threats are sea temperature rise, sea level rise and pH changes from ocean acidification, all associated with greenhouse gas emissions. A 2014 study lists factors such as population explosion along the coast lines, overfishing, the pollution of coastal areas, global warming and invasive species among the main reasons that have put reefs in danger of extinction.

A study released in April 2013 has shown that air pollution can also stunt the growth of coral reefs; researchers from Australia, Panama and the UK used coral records (between 1880 and 2000) from the western Caribbean to show the threat of factors such as coal-burning coal and volcanic eruptions. Pollutants, such as Tributyltin, a biocide released into water from in anti-fouling paint can be toxic to corals.

In 2011, researchers suggested that "extant marine invertebrates face the same synergistic effects of multiple stressors" that occurred during the end-Permian extinction, and that genera "with poorly buffered respiratory physiology and calcareous shells", such as corals, were particularly vulnerable.

Rock coral on seamounts across the ocean are under fire from bottom trawling. Reportedly up to 50% of the catch is rock coral, and the practice transforms coral structures to rubble. With it taking years to regrow, these coral communities are disappearing faster than they can sustain themselves.

Another cause for the death of coral reefs is bioerosion. Various fishes graze corals, dead or alive and change the morphology of coral reefs making them more susceptible to other physical and chemical threats. It has been generally observed that only the algae growing on dead corals is eaten and the live ones are not. However, this act still destroys the top layer of coral substrate and makes it harder for the reefs to sustain.

In El Niño-year 2010, preliminary reports show global coral bleaching reached its worst level since another El Niño year, 1998, when 16% of the world's reefs died as a result of increased water temperature. In Indonesia's Aceh province, surveys showed some 80% of bleached corals died. Scientists do not yet understand the long-term impacts of coral bleaching, but they do know that bleaching leaves corals vulnerable to disease, stunts their growth, and affects their reproduction, while severe bleaching kills them. In July, Malaysia closed several dive sites where virtually all the corals were damaged by bleaching.

To find answers for these problems, researchers study the various factors that impact reefs. The list includes the ocean's role as a carbon dioxide sink, atmospheric changes, ultraviolet light, ocean acidification, viruses, impacts of dust storms carrying agents to far-flung reefs, pollutants, algal blooms and others. Reefs are threatened well beyond coastal areas. Coral reefs with one type of zooxanthellae are more prone to bleaching than are reefs with another, more hardy, species.

General estimates show approximately 10% of the world's coral reefs are dead. About 60% of the world's reefs are at risk due to destructive, human-related activities. The threat to the health of reefs is particularly high in Southeast Asia, where 95% of reefs are at risk from local threats. By the 2030s, 90% of reefs are expected to be at risk from both human activities and climate change; by 2050, all coral reefs will be in danger.

Current research is showing that ecotourism in the Great Barrier Reef is contributing to coral disease, and that chemicals in sunscreens may contribute to the impact of viruses on zooxanthellae.

Protection

Marine protected areas (MPAs) have become increasingly prominent for reef management. MPAs promote responsible fishery management and habitat protection. Much like national parks and wildlife refuges, and to varying degrees, MPAs restrict potentially damaging activities. MPAs encompass both social and biological objectives, including reef restoration, aesthetics, biodiversity, and economic benefits. However, there are very few MPAs that have actually made a substantial difference. Research in Indonesia, Philippines and Papua New Guinea shows that there is no significant difference between an MPA site and an unprotected site. Conflicts surrounding MPAs involve lack of participation, clashing views of the government and fisheries, effectiveness of the area, and funding. In some situations, as in the Phoenix Islands Protected Area, MPAs can also provide revenue, potentially equal to the income they would have generated without controls, as Kiribati did for its Phoenix Islands.

A diversity of corals.

According to the *Caribbean Coral Reefs - Status Report 1970-2012* made by the IUCN. States that; stopping overfishing especially key fishes to coral reef like parrotfish, coastal zone management which reduce human pressure on reef, (for example restricting the coastal settlement, development and tourism in coastal reef) and controlling pollution specially sewage wastage, may not only reduce coral declining but also reverse it and may let to coral reef more adaptable to changes relates to climate and acidification. The report shows that healthier reef in the Caribbean are those with large population of parrotfish in countries which protect these key fishes and sea urchins, banning fish trap and Spearfishing creating "resilient reefs".

To help combat ocean acidification, some laws are in place to reduce greenhouse gases such as carbon dioxide. The Clean Water Act puts pressure on state government agencies to monitor and limit runoff of pollutants that can cause ocean acidification. Stormwater surge preventions are also in place, as well as coastal buffers between agricultural land and the coastline. This act also ensures that delicate watershed ecosystems are intact, such as wetlands. The Clean Water Act is funded by the federal government, and is monitored by various watershed groups. Many land use laws aim to reduce CO_2 emissions by limiting deforestation. Deforestation causes erosion, which releases a large amount of carbon stored in the soil, which then flows into the ocean, contributing to ocean acidification. Incentives are used to reduce miles traveled by vehicles, which reduces the carbon emissions into the atmosphere, thereby reducing the amount of dissolved CO_2 in the ocean. State and federal governments also control coastal erosion, which releases stored carbon in the soil into the ocean, increasing ocean acidification. High-end satellite technology is increasingly being employed to monitor coral reef conditions.

Biosphere reserve, marine park, national monument and world heritage status can protect reefs. For example, Belize's barrier reef, Chagos archipelago, Sian Ka'an, the Galapagos islands, Great Barrier Reef, Henderson Island, Palau and Papahānaumokuākea Marine National Monument are world heritage sites.

In Australia, the Great Barrier Reef is protected by the Great Barrier Reef Marine Park Authority, and is the subject of much legislation, including a biodiversity action plan. They have compiled a Coral Reef Resilience Action Plan. This detailed action plan consists of numerous adaptive management strategies, including reducing our carbon footprint, which would ultimately reduce the amount of ocean acidification in the oceans surrounding the Great Barrier Reef. An extensive public awareness plan is also in place to provide education on the "rainforests of the sea" and how people can reduce carbon emissions, thereby reducing ocean acidification.

Inhabitants of Ahus Island, Manus Province, Papua New Guinea, have followed a generations-old practice of restricting fishing in six areas of their reef lagoon. Their cultural traditions allow line fishing, but no net or spear fishing. The result is both the biomass and individual fish sizes are significantly larger than in places where fishing is unrestricted.

Restoration

Coral fragments growing on nontoxic concrete.

Coral aquaculture, also known as coral farming or coral gardening, is showing promise as a potentially effective tool for restoring coral reefs, which have been declining around the world. The process bypasses the early growth stages of corals when they are most at risk of dying. Coral seeds are grown in nurseries, then replanted on the reef. Coral is farmed by coral farmers who live locally to the reefs and farm for reef conservation or for income.

Efforts to expand the size and number of coral reefs generally involve supplying substrate to allow more corals to find a home. Substrate materials include discarded vehicle tires, scuttled ships, subway cars, and formed concrete, such as reef balls. Reefs also grow unaided on marine structures such as oil rigs. In large restoration projects, propagated hermatypic coral on substrate can be secured with metal pins, superglue or milliput. Needle and thread can also attach A-hermatype coral to substrate.

A substrate for growing corals referred to as Biorock is produced by running low voltage electrical currents through seawater to crystallize dissolved minerals onto steel structures. The resultant white carbonate (aragonite) is the same mineral that makes up natural coral reefs. Corals rapidly colonize and grow at accelerated rates on these coated structures. The electrical currents also accelerate formation and growth of both chemical limestone rock and the skeletons of corals and other shell-bearing organisms. The vicinity of the anode and cathode provides a high-pH environment which inhibits the growth of competitive filamentous and fleshy algae. The increased growth rates fully depend on the accretion activity.

During accretion, the settled corals display an increased growth rate, size and density, but after the process is complete, growth rate and density return to levels comparable to natural growth, and are about the same size or slightly smaller.

One case study with coral reef restoration was conducted on the island of Oahu in Hawaii. The University of Hawaii has come up with a Coral Reef Assessment and Monitoring Program to help relocate and restore coral reefs in Hawaii. A boat channel on the island of Oahu to the Hawaii Institute of Marine Biology was overcrowded with coral reefs. Also, many areas of coral reef patches in the channel had been damaged from past dredging in the channel. Dredging covers the existing corals with sand, and their larvae cannot build and thrive on sand; they can only build on to existing reefs. Because of this, the University of Hawaii decided to relocate some of the coral reef to a different transplant site. They transplanted them with the help of the United States Army divers, to a relocation site relatively close to the channel. They observed very little, if any, damage occurred to any of the colonies while they were being transported, and no mortality of coral reefs has been observed on the new transplant site, but they will be continuing to monitor the new transplant site to see how potential environmental impacts (i.e. ocean acidification) will harm the overall reef mortality rate. While trying to attach the coral to the new transplant site, they found the coral placed on hard rock is growing considerably well, and coral was even growing on the wires that attached the transplant corals to the transplant site. This gives new hope to future research on coral reef transplant sites. As a result of this coral restoration project, no environmental effects were seen from the transplantation process, no recreational activities were decreased, and no scenic areas were affected by the project. This is a great example that coral transplantation and restoration can work and thrive under the right conditions, which means there may be hope for other damaged coral reefs.

Another possibility for coral restoration is gene therapy. Through infecting coral with genetically modified bacteria, it may be possible to grow corals that are more resistant to climate change and other threats.

Reefs in the Past

Throughout Earth history, from a few thousand years after hard skeletons were developed by marine organisms, there were almost always reefs. The times of maximum development were in the

Middle Cambrian (513–501 Ma), Devonian (416–359 Ma) and Carboniferous (359–299 Ma), owing to order Rugosa extinct corals, and Late Cretaceous (100–66 Ma) and all Neogene (23 Ma–present), owing to order Scleractinia corals.

Ancient coral reefs.

Not all reefs in the past were formed by corals: those in the Early Cambrian (542–513 Ma) resulted from calcareous algae and archaeocyathids (small animals with conical shape, probably related to sponges) and in the Late Cretaceous (100–66 Ma), when there also existed reefs formed by a group of bivalves called rudists; one of the valves formed the main conical structure and the other, much smaller valve acted as a cap.

Measurements of the oxygen isotopic composition of the aragonitic skeleton of coral reefs, such as Porites, can indicate changes in the sea surface temperature and sea surface salinity conditions of the ocean during the growth of the coral. This technique is often used by climate scientists to infer the paleoclimate of a region.

Chapter 6

Marine Microbes

Marine microbes are tiny, single celled organisms that live in the oceans and make up a majority of the ocean biomass. They consist of all sorts of bacteria, viruses and fungi. Marine microbiology is the study of these organisms, their biodiversity, ecology and biogeochemistry. This chapter has been carefully written to provide an easy understanding of marine microbes.

Marine microbes are tiny organisms that live in marine environments and can only be seen under a microscope. They include cellular life forms, such as bacteria, fungi, algae and plankton, along with the viruses that freeload along with them.

More than a billion microorganisms live in each litre of seawater—they dominate the abundance, diversity and metabolic activity of the ocean.

Microbes comprise 98 per cent of the biomass of the world's oceans, supply more than half the world's oxygen and are the major processors of the world's greenhouse gases, which means they have the potential to mitigate the effects of climate change.

Scientists are only just beginning to understand the important environmental roles that microbes play in marine systems—from feeding ecosystems to consuming waste and sequestering carbon. Scientists are investigating several areas where microbial processes are central to issues of immediate concern to the world's coral reefs, including:

- Indicators of ecosystem health and environmental impacts.
- Adaptation, acclimatization and evolution of coral reef organisms in the face of global change.
- Marine microbiomes and viromes (the ecological communities of microbes and viruses, respectively).

A surprising number of these marine microbes have only recently been discovered, or are known only by their DNA. In some cases, we only know that they are really tiny and have certain types of genes, such as genes necessary for fixing nitrogen. This is one reason that there are no photographs of many of these organisms.

At the broadest level, these microbes can be divided up between three main groups, which are three domains of living things on this planet:

- Bacteria

- Archaea, which look similar to bacteria, but are an entirely separate domain.
- Eukaryotes, a group that includes animals, plants, fungi, protists, and algae.

The microbes listed below are grouped by these three domains. Within each domain, they are grouped according to their roles in nitrogen cycling. We have also included a few microbes that are key players in open-ocean photosynthesis and primary production. These microbes are important for nitrogen cycling in that they "assimilate" nitrogen and convert it into living tissue or "particulate nitrogen."

Several terms are used to describe microbes that perform certain roles in nitrogen cycling, but which may or may not be related to one another. The bacteria and archaea that can convert nitrogen gas into more biologically useful forms such as ammonium are known as "nitrogen fixers" or "diazotrophs." Microbes that convert ammonium to nitrate are called "nitrifying organisms."

Marine Viruses

Viruses in the Sea

Viruses can be found wherever life is found and in seawater they outnumber Bacteria, Archaea, and Eukarya combined by an order of magnitude or more (Wommack and Colwell, 2000;Suttle, 2007). The concentration of virions in surface seawater in the open ocean is on the order of 5–10 billion per liter. Their concentration declines with depth, but viruses are present in the deepest waters sampled (>4000 m). The concentration of virions in surface marine sediments is typically even higher than in the immediate overlying waters. As in the water column, viruses become less abundant with depth in the sediment, but remain detectable to depths >100 m below the seafloor and in sediments with estimated ages of hundreds of thousands of years or more.

Virions suspended in seawater are a component of the marine plankton and are therefore often referred to as virioplankton. Those in sediments are sometimes referred to as viriobenthos. Plankton ecologists find it convenient at times to refer to plankton according to which of seven different size categories they belong. Because of their small size, most of the viruses in seawater are members of the smallest named size category, the femtoplankton (0.02–0.2 μm), but larger viruses are also part of the picoplankton (0.2–2 mm), a size class that also includes almost all of the bacteria and archaea and the smallest protists.

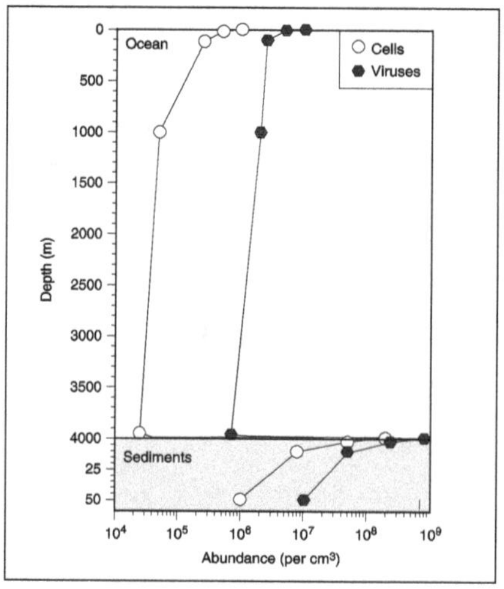

Idealized plot of the concentrations of viruses and cells in open ocean seawater and sediments.

Sediments are indicated by the shaded area. Sediment depth (in bold) starts over from zero at the sediment–water interface and is plotted at a different scale for clarity. "Cells" refers to all cells, but these numbers are dominated by bacteria and archaea. The numbers shown are approximate and concentrations at a given depth can vary by an order of magnitude depending on location. Concentrations tend to be higher in areas of high productivity such as the coastal zone.

Viral infections have been reported for a broad range of marine organisms and these infections have wide-ranging effects on everything from evolution, ecology, and ocean biogeochemistry to human health and economics. Viral infections of phytoplankton, the foundation of the marine food web, affect how much of the ocean's primary production is transferred to higher trophic levels. Large increases in phytoplankton concentrations (blooms) are a common phenomenon in the ocean and these periodic in-creases in primary production serve as a critical food supply for life in the sea. Viruses also spread more easily through populations at high density, however, and phytoplankton blooms can be decimated by the rapid spread of viral infection. Lysis of phytoplankton decreases the efficiency with which they are consumed by grazers and much of the primary production is instead decomposed by bacteria. Bacteria are themselves susceptible to viral infections and viral lysis of bacteria likewise diverts material and energy from protistan grazers, which favors further growth and respiration by other uninfected bacteria. It thus appears that the net effect of lytic viral activity in the marine food web is to decrease the amount of carbon that reaches higher trophic levels (e.g., fish) while increasing the amount of carbon, which is respired by the bacteria.

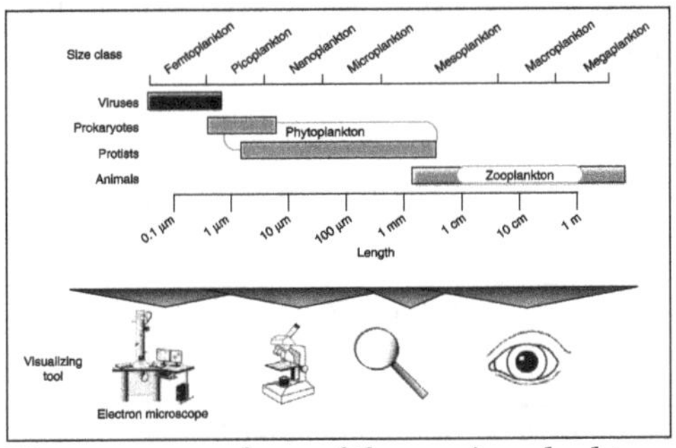

Viruses as members of the marine plankton.

As a group, viruses are the smallest and most abundant members of the plankton and an electron microscope is required to reveal their details. Plankton are often named by the size category to which they belong, but may also be referred to in functional categories such as virioplankton, phytoplankton, and zooplankton.

From the perspective of humans, viruses could be lamented as reducing the efficiency of wild fisheries, but by how much is unknown. In any case, there is little to be done about the persistent, ubiquitous influence of viruses on the ocean's food web; it is an inescapable feature of the natural cycle of life in the sea. However, viruses are a legitimate human concern with tangible economic consequences when they cause death and disease among farmed marine species such as shrimp and fish. As wild fisheries become increasingly threatened by overexploitation, humans are relying more heavily on aqua- and mariculture as a source of protein. Intensive mariculture involves growth of single species at high densities, which, unfortunately, is a condition under which viral and other diseases can easily spread within the farm and from the farm to wild populations. The economic losses resulting from viral diseases in aquaculture are estimated to be in the billions of dollars annually so there is a strong incentive to find solutions to combat viral diseases in commercially exploited marine species.

Life in the sea is very diverse, with representation of most of the major types of life on the planet. As one might then expect, the diversity of viruses in the sea is also broad. Among viruses infecting the marine organisms, members of all Baltimore groups except for Group VII have been found so far.

Taxonomic Viruses

Specific genera within a family are noted in some cases where it seems of particular interest. The taxonomic groups represented are, for the most part, restricted to those explicitly mentioned in the most recent report of the ICTV, but some information is included from other sources and proposals that have been accepted since the report was completed. For the purposes of this presentation, a virus was considered marine if it infects an organism commonly referred to as marine. This includes organisms for which the entire life cycle is completed in the ocean, but also sea turtles (marine reptile), a variety of birds such as gulls, terns, cormorants, and pelicans (marine birds) as well as seals and sea lions (marine mammals). The general characteristics of the virions in each taxon and the types of marine organisms they infect are noted.

Group	*Genome*		*Taxonomy*			*Marine hosts*	
	Type	*Segments*	*Order*	*Family*	*Subfamily*	*Types*	*Names*
I	dsDNA	Mono	*Caudovirales*	*Myoviridae*	–	B	Cyanobacteria and proteobacteria
				Podoviridae	–	B	Cyanobacteria and proteobacteria
				Siphoviridae	–	B	Cyanobacteria and proteobacteria
			Herpesvirales	*Alloherpesviridae*	–	V	Fish
				Herpesviridae	*Alphaherpesvirinae*	V	Turtle
					Gammaherpesvirinae	V	Seal
				Malacoherpesviridae	–	I	Oyster
			Unassigned	*Adenoviridae*	–	V	Fish
				Corticoviridae	–	B	Bacteria
				Iridoviridae	–	V	Fish
				Mimiviridae	–	Pr	Heterotrophic protist
				Nimaviridae	–	I	Shrimp and crab
				Papillomaviridae	–	V	Porpoise and dolphin
				Phycodnaviridae	–	Pr	Micro- and macroalga
				Poxviridae	–	V	Marine mammals and birds
				Unassigned	–	Pr	Microalga

II	ssDNA	Mono	Unassigned	*Anelloviridae*	–	V	Sea lion
				Circoviridae	–	V	Herring gull
				Inoviridae	–	B	Proteobacteria
				Microviridae	*Gokushovir-inae*	B	Proteobacteria
				Parvoviridae	–	I	Shrimp
				Unassigned	–	Pr	Microalga
III	dsRNA	Bi	Unassigned	*Birnaviridae*	–	I, V	Molluscs, crus-taceans, and fish
		Multi	Unassigned	*Reoviridae*	*Spinareovir-inae*	I, V	Bivalves and fish
					Sedoreoviri-nae	Pr, I	Microalga and crab
IV	ssRNA (+)	Mono	*Nidovirales*	*Roniviridae*	–	I	Shrimp
		Mono	*Picornavi-rales*	*Dicistroviridae*	–	I	Shrimp
				Marnaviridae	–	Pr	Microalga
				Unassigned	–	Pr	Labrynthulid and microalga
		Mono	Unassigned	*Astroviridae*	–	V	Sea lion and dolphin
				Caliciviridae	–	V	Seals and whales
		Bi		*Nodaviridae*	–	V	Fish
		Mono		*Togaviridae*	–	V	Fish and mam-mals
V	ssRNA (-)	Mono	*Mononegavi-rales*	*Paramyxoviri-dae Rhabdovi-ridae*	*Paramyxovir-inae* –	V V	Seals and whales Fish
		Multi	Unassigned	*Orthomyxoviridae*	–	V	Fish, sea birds, seals, and whales
VI	ssRNA (+)	Mono	Unassigned	*Metaviridae*	–	I, V	Sea urchin and fish

Group I: dsDNA Viruses

The dsDNA virus isolates known to infect marine organisms are assigned to 13 of the 30 families in Baltimore Group I. Three families are members of the order Caudovirales, another three are members of the order Herpesvirales, and six families are not assigned to an order. A number of viruses infecting marine plankton have also not yet been assigned to a family.

Order Caudovirales

The viral order Caudovirales contains three related families of tailed viruses, all of which infect bacteria or archaea. Members of this order have nonenveloped capsids with icosahedral symmetry, but these may be isometric or prolate with diameters ranging from around 40–145 nm in diameter. Helical tails (10–355 nm long and 5–18 nm wide) aid in attachment to the host cell and serve as a conduit through which the viral genome is delivered to the interior of the cell. The tail structure serves as a taxonomic character that divides the order into the families Myoviridae, Podoviridae, and Siphoviridae can be found in all three of these families, but marine isolates reported to date are restricted to those infecting bacteria, probably because isolates of marine archaea remain so rare. Members of all three families are very common in seawater.

Family Myoviridae

Viruses in this family are characterized by having a long (50 to more than 200 nm) hollow tail with a neck and a contractile sheath. Because of the sheath, the tails of these viruses are thicker (16–18 nm) than those of other caudovirads. Myovirids that infect marine bacteria, particularly bacteria in the phylum Gammaproteobacteria, have been classified, but many more marine isolates are known. These include a large number of isolates infecting cyanobacteria in the genera Prochlorococcus, and Synechococcus (Sullivan et al., 2010). Although there is considerable overlap, the genomes and capsids of myovirids tend to be larger than those of the other two families in this order.

Family Podoviridae

Viruses in this family are characterized by the presence of a short, noncontractile tail that is 10–48 nm long and about 8 nm wide. Podovirids appear to be the most numerous of all tailed phages in the marine environment. Podovirids infecting marine bacteria in the phyla Alpha- and Gammaprotoebacteria have been officially classified by the ICTV, but many more isolates infecting a range of bacteria exist, including podovirids infecting Prochlorococcus and Synecho- coccus.

Family Siphoviridae

Viruses in this family have a long, noncontractile tail that is around 100 to > 4300 nm long, and 5–10 nm wide. These long, unsheathed tails often appear to be quite flexible, displaying bends and curves in electron micrographs.

Order Herpesvirales

Members of the order Herpesvirales have a large icosahedral capsid enclosed in a tegument and a lipid bilayer membrane. The genome ranges from 125,000 to 295,000 bp. Capsids are around 125 nm, but the entire virion is closer to 200 nm in diameter. These viruses are very wide- spread in the vertebrates. Examples of herpesvirads infecting marine vertebrates are known in all three families of the order.

Family Alloherpesviridae

Members of this family include those that infect fish and amphibians. Marine and estuarine representatives are in the genus Ictaluvirus and include those that infect catfish, sturgeon, salmon, and cod.

Family Herpesviridae

This family consists of genetically related herpesvirids that infect reptiles, birds, and mammals. Marine representatives have been assigned to two of the three subfamilies (Alpha- and Gammaherepesvirinae), but are not yet assigned to a genus. Marine hosts include sea turtles, seals, sea lions, dolphins, and whales.

Family Malacoherpesviridae

This new family consists of a single genus (Ostreavirus) and species (Ostreid herpesvirus 1). This virus infects oysters and is the only herpesvirad identified so far that infects an invertebrate.

Families not Assigned to an Order

Family Adenoviridae

Members of this family are nonenveloped and have a capsid 70–90 nm in diameter with icosahedral symmetry and a genome between 25,000 and 50,000 bp. Adenovirids infect a wide variety of vertebrates. There are a number of reports of infections by adenovirids in marine mammals (pinnipeds and cetaceans) and in gulls, but none have been isolated and officially classified. Sea lions may be a host for canine adenovirus. The only reported fish adenovirus is in the genus Ichtadenovirus with the sole representative being Sturgeon ade- novirus A, which infects white sturgeon. Although not truly marine, the sturgeon habitat includes brackish coastal waters.

Family Corticoviridae

This family is represented by a single genus (Corticovirus) and a single species Pseudoalteromonas phage PM2) that infects a marine bacterium. The virion has an icosahedral capsid 57 nm in diameter, no tail, and a lipid bilayer membrane interior to the capsid. The genome in the virion is supercoiled with a length of 10,079 bp. Although only one species has been classified at present, related sequences have been detected within the genomes of a variety of bacteria suggesting that they could be prevalent in the marine environment.

Family Iridoviridae

Members of this family have a large icosahedral capsid (120–200 nm diameter) and a large genome (around 100,000 to 200,000 bp). They infect invertebrates (arthropods) and vertebrates (amphibians, reptiles, and fish). Those infecting invertebrates are nonenveloped while those infecting vertebrates acquire a lipid envelope that derives from the host cell membrane during virion budding. Members of two of the five genera in this family (Lymphocystivirus and Megalocytivirus) infect marine fish.

Family Mimiviridae

This is a relatively new family of viruses, the members of which have unusually large genomes and icosahedral capsids. The first representative to be fully sequenced, Acanthamoeba polyphaga mimivirus (APMV) infects a freshwater amoeba. At the time of its discovery, it was the largest known virus with a genome of 1,181,549 bp and a capsid diameter of 500 nm. The virus is densely covered with hair-like fibrils that extend its effective diameter to 750 nm. Recently, another virus related to APMV with a larger genome (1,259,157 bp), but smaller capsid (440 nm) was isolated from a marine sample using several freshwater Acanthamoeba species as hosts. The natural host of this new isolate, named Megavirus chilensis, is unknown, but is presumed to be a marine or estuarine protist. This is the largest viral genome known at the present time.

Family Nimaviridae

This family consists at present of a single genus (Whispovirus) and species (White spot syndrome virus). Virions consist of a rod- shaped nucleocapsid (approximately 65–70 x 300–350 nm) enclosed in a tegument and a trilaminar envelope. The overall virion dimensions are 120–150 nm in diameter and 270– 290 nm in length. The virion is unusual in having a thread-like appendage that gives it the superficial appearance of a bacterium with a polar flagellum.

Family Phycodnaviridae

The members of this family that have been officially classified so far have large icosahedral capsids (around 100–200 nm diameter) with no outer envelope and genome sizes ranging from approximately 100,000 to 500,000 bp. These viruses infect photosynthetic protists (micro and macroalgae) in both freshwater (genus Chlorovirus) and the marine environment (genera Coccolithovirus, Phaeovirus, Prasinoviruses, Prymnesio- viruses, and Raphidoviruses). Members of the genus Phaeovirus infect filamentous brown algae, but all others infect unicellular plankton.

Family Poxviridae

The capsids of poxvirids are enveloped and brick shaped or ovoid with tubular surface processes. They vary from 200 to 450 nm in length, 140–280 nm in width, and 140–280 nm in thickness. The genomes are linear dsDNA ranging from 130,000 to 380,000 bp. Members of this family infect a wide range of hosts. The poxvirids infecting vertebrates are in the subfamily Chordopoxvirinae, which contains nine genera. The subfamily Entomopoxivirinae comprises three genera, the members of which infect various insects. The family is well known for the serious human disease caused by one of its notorious members, small pox (now eradicated). In the marine realm, poxviruses have been identified in penguins, seals, sea lions, dolphins, and whales, but none of these have been officially classified.

Unassigned

Several new viruses infecting marine heterotrophic and phototrophic protists have been isolated and described, for which the taxonomic positions remain unclear. A large (300 nm) virus, CroV, that infects Cafeteria roenbergensis is the first isolate that infects a phagotrophic marine protist. Based on the large genome size (750,000 bp) and specific sequence of the DNA polymerase of CroV, it may belong to the family Mimiviridae. A number of other virus isolates infecting photosynthetic marine protists (Chrysochromulina ericina virus 01B, Phaeocystis pouchetii virus 01, and Pyramimonas orientalis virus 01B) have capsids closer in size to phycodnavirids (140–220 nm), but their DNA polymerase genes place them closer to APMV and CroV, suggesting they may also be members of the Mimiviridae.

A large virus (200 nm diameter, 350,000 bp genome) was isolated that infects the bivalve-killing dinoflagellate, Heterocapsa circularisquama . The DNA poly- merase sequence of this isolate is very similar to that of African swine fever viruses in the genus Asfivirus, which is the sole genus in the family Asfarviridae. It is, therefore, highly di- vergent from other dsDNA viruses infecting phytoplankton and a new genus name, Dinodnavirus, has been proposed for this virus.

Group II: ssDNA Viruses

Family Anelloviridae

Members of this family have a nonenveloped icosahedral capsid approximately 30 nm in size and a circular genome ranging in size from 2000 to 3900 nt. They appear to be widespread in mammals, but their involvement in specific diseases remains unproven. Novel anellovirids have recently been detected by metagenomic analyses of lung tissue from seals and sea lions, but these have yet to be approved as species.

Family Circoviridae

This family is comprised of two genera (Circovirus and Gyro virus) known to infect birds and swine, although distantly related viruses have been described in primates. The virions are nonenveloped and appear spherical with icosahedral symmetry. The capsid diameter is 20.5–25 nm and the gen- ome ranges from 1775 to 2300 nt. The only classified member so far that might be considered marine is the gull circovirus in the genus Circovirus.

Family Inoviridae

Virions in this family are filaments or rods with helical symmetry with a circular genome ranging from 4500 to 12,400 nt in length. Those in the genus Inovirus are long (700–900 nm) and thin (7 nm) and infect gammaproteobacteria, while those in the genus Plectrovirus (10–15 nm wide and 70–280 nm long) infect Gram-positive bacteria. Among the known hosts of inovirids are members of the bacterial genus Vibrio, which are common in coastal estuarine and marine waters. Inovirids cause persistent infections and are responsible for lysogenic conversion. The inovirid, Vibrio cholera CTXø, for example, codes for an enterotoxin that converts Vibrio cholerae into a human pathogen.

Family Microviridae

Virions in this family have a small, nonenveloped capsid ap- proximately 25–27 nm in diameter and a circular, positive- sense genome ranging in size from 5300 to 6100 nt (genus Microvirus) or from 4400 to 4900 nt (genera Chlamydiamicrovirus, Bdellomicrovirus, and Spiromicrovirus in the subfamily Gokushovirinae). The hosts for members of the genus Bdellomicrovirus, are bacteria in the genus Bdellovibrio. Bdellovibrio spp. are predatory bacteria found in a range of freshwater, estuarine, and marine habitats. All strains of bdellomicroviruses isolated so far have been from freshwater. However, there is evidence from metagenomics that additional members of the Gokushovirinae are widespread in the ocean. Sequence comparisons indicate that these viruses are distantly related to known chlamydiamicroviruses and bdellomicroviruses, but their hosts have not yet been identified.

Unassigned

Several genetically similar ssDNA viruses infecting diatoms in the genus Chaetoceros have been isolated. The virions are nonenveloped and have an icosahedral capsid ranging in size from 30 to 38 nm and a genome around 6000 nt in length. Although these viruses do not yet belong to an official family, a new genus, Bacilladnavirus, has been approved to accommodate this group of viruses.

Group III: dsRNA Viruses

Family Birnaviridae

There are four genera in the family Birnaviridae. The icosahedral particles of birnavirids are 65 nm in diameter, none- nveloped and are composed of a single capsid protein. The genome consists of two linear segments of dsRNA approximately 5900–6900 bp in total length that encodes a poly- merase and a large structural polyprotein that is cleaved post-translationally. Birnavirids infect rotifers, insects, fish, and birds and are globally distributed. In the marine en- vironment, aquabirnaviruses cause disease in salmonids, amberjack (yellowtail), bream, and cod among other fishes, and even molluscs and crustaceans.

Family Reoviridae

The family Reoviridae has fifteen genera divided among two subfamilies, the Spinareovirinae that include viruses that have turret-like proteins projecting from their capsid, and the Sedoreovirinae, the reovirids without turrets. The virions in this family are nonenveloped icosahedrons that range from 49 to 100 nm in diameter. Reovirids are distinctive in that their capsid can be composed of as many as three concentric protein shells that surround from 9 to 12 segments of dsRNA, that together range between 18,500 and 33,300 bp in total length. The viruses in this family infect a wide range of multicellular organisms including protists, fungi, plants, insects, reptiles, and humans. The dsRNA virus Micromonas pusilla reovirus (MpRV) from the genus Mimireovirus was one of the first marine RNA viruses isolated and it infects a marine protist . The host, M. pusilla, is a small (<2 mm), autotrophic flagellate considered the most abundant picoeukaryote in the global oceans. Species of bivalves and finfish are subject to infection from reovirids in the genus Aquareovirus, while orbiviruses have been found in seabirds (e.g., murres, puffins, and razorbills) and cardoreoviruses in crabs.

Group IV: Positive-Sense ssRNA Viruses

Order Nidovirales

The order Nidovirales is divided into three families, two sub- families and seven genera. Virions in this order range widely in size from 50 to 200 nm, but all are enveloped. The genome of a nidovirad is a single, linear molecule of RNA that ranges in size from 12,700 to 29,100 bp and has a 50 cap and a poly(A) tail that extends from the 30 end. The gene order is highly conserved, with open reading frames (ORFs) nearest the 50 end encoding the replicase, followed by 3–12 structural proteins. The genomes of viruses in the Nidovirales contain a characteristic - 1 frameshift that is involved in the regulation of viral protein production during replication. Members of this order are the largest RNA viruses known to date, both in genome and particle size.

Family Roniviridae

The genus Okavirus is the only genus in the family Roniviridae. The virions of ronivirids are enveloped, bacilliform in shape, vary in size from 150 to 200 nm, and are studded with spike-like glycoprotein projections that have bulbous distal ends. The genome of these viruses has features typical of the order Nidovirales and can be quite large for an RNA virus, exceeding

26,000 nt in length. The genome configuration, particularly in the structural genes, as well as the absence of a discrete structural membrane protein gene, distinguishes ronivirads from other viruses in the order. Viruses in the family Roniviridae infect the giant tiger shrimp and the blue shrimp.

Order Picornavirales

The sprawling order of the Picornavirales contains five families, two subfamilies, 24 genera, and approximately 120 species of viruses. Picornavirads are "picorna-like" in that they have similar features to viruses in the family Picornaviridae. Members of the order have a nonenveloped icosahedral capsid that ranges in diameter from 22 to 30 nm. The linear, ssRNA genome is 7000–15,500 nt in length and encodes a polyprotein that is post-translationally processed. Picornavirads also have in common a conserved nonstructural gene order and a small protein covalently linked to the 50 end and a 30 end poly(A) tail. These viruses infect a wide range of eukaryotes, including humans. Picornavirads of note include human enterovirus C, which causes poliomyelitis, human rhinovirus, responsible for the common cold, and Israeli acute paralysis virus, which has been linked to honey bee colony collapse disorder.

Family Dicistroviridae

The family Dicistroviridae has two recognized genera, the cri- paviruses and the aparaviruses. The particles of dicistrovirids are nonenveloped icosahedra approximately 30 nm in size. The genome is a single molecule of linear RNA that can be from 8500 to 10,000 nt in length and encodes two poly- proteins, or cistrons (ergo the name of the family). Dicistrovirids infect only invertebrates, including aphids, bees, and butterflies. In the marine environment, Taura syndrome virus (TSV), an aparavirus, infects penaeid shrimp. TSV epidemics have been responsible for high rates of mortality in shrimp farms.

Family Marnaviridae

Heterosigma akashiwo RNA virus (HaRNAV) is the only member of the Marnaviridae. Heterosigma akashiwo, the host of HaRNAV, is a small, motile photosynthetic protist that is a common member of the temperate coastal phytoplankton community. During times when surface waters are stratified as a result of higher sea surface temperatures and low winds, Heterosigma akashiwo is capable of forming immense blooms that can cause extensive fish kills. These toxic blooms affect the fish farming industry in the fjords of western Canada, and Scandanavia in particular. In 2003, Tai and colleagues isolated HaRNAV, the first RNA virus reported to infect a protist . HaRNAV has a nonenveloped, icosahedral capsid that is approximately 25 nm in diameter. The burst size of HaRNAV is estimated to be 20,000 viruses per cell. The linear ssRNA genome of HaRNAV is 8600 nt in length and is monocistronic – unlike AuRNAV, the bacillar- naviruses, TSV, and HcRNAV. However, the HaRNAV genome has the same gene order as the other marine ssRNA viruses with the nonstructural genes nearest the 50 end of the genome followed by the structural genes. Phylogenies based on the conserved domain sequences from HaRNAV and the corresponding regions of other viruses place HaRNAV outside previously characterized virus taxa within the order Picornavirales.

Proposed Genus Bacillarnavirus

The genus Bacillarnavirus has not been assigned to a family within the order Picornavirales. It has three recognized mem- bers thus far, Rhizosolenia setigera RNA virus (RsetRNAV), Chaetoceros tenuissimus Meunier RNA virus (CtenRNAV), and Chaetoceros socialis f. radians RNA virus (CsfrRNAV) all of which infect marine diatoms. Diatoms are widely distributed, highly diverse, and are among the most abundant eukaryotes in the ocean. They form a large fraction of the total marine primary producers and are thus significant contributors to the cycling of energy in the ocean. The hosts of CtenRNAV and CsfrRNAV are members of the genus Chaetoceros, a group of chain-forming diatoms that can form blooms associated with fish kills. The two viruses have icosahedral symmetry and have ssRNA genomes that are dicistronic and approximately 9500 nt in length. CtenRNAV has an estimated capsid diameter of 31 nm, while that of CsfrRNAV is only 22 nm. A striking difference between the two viruses is that CtenRNAV has an estimated burst size of 10,000 virions while CsfrRNAV produces approximately 66 per lytic event. RsetRNAV was the first virus brought into culture that infects a diatom. Like CtenRNAV and CsfrRNAV, RsetRNAV has an icosahedral capsid (approximately 32 nm in diameter), replicates in the cyto- plasm and has a positive-sense ssRNA dicistronic genome ap- proximately 8900 long. The genomes of all three RNA viruses have a similar gene order, with the structural genes nearest the 50 end, followed by the helicase, protease, and replicase and terminating in a poly(A) tail. Alignments of the replicase genes of these viruses clearly demonstrate that they are closely related and form a well- supported monophyletic clade.

Proposed Family Labyrnaviridae

Aurantiochytrium RNA virus (AuRNAV) is the only member of the family Labyrnaviridae. AuRNAV infects the thraustochytrid Aurantiochytrium sp. Thraustochytrids are single-celled marine organisms that were once thought to be fungi, but are now recognized as a class of protists. These organism are capable of feeding directly through absorption and are involved in the degradation of detritus. Like HcRNAV and the bacillarnaviruses, AuRNAV has a small (roughly 25 nm), nonenveloped icosa- hedral capsid. AuRNAV replicates in the cytoplasm and has a phenomenal burst size of approximately 60,000 particles per lytic event. The genome of AuRNAV is a single molecule of RNA 9000 nt in length with a poly(A) tail. The three ORFs of the AuRNAV genome encode the capsid proteins, nonstructural genes and a protein of unknown function. Interestingly, AuRNAV produces subgenomic RNA during replication. Although the specific mechanism is unknown, this suggests that the AuRNAV genome does not possess an internal ribosome entry site structure that is characteristic of other picornavirads.

Families Not Assigned to an Order

Proposed Family Alvernaviridae

The family Alvernaviridae has a single representative at present, Heterocapsa circularisquama RNA virus (HcRNAV), which infects the eponymous dinoflagellate. The dinoflagellates are a diverse group of unicellular flagellates that are capable of de- riving energy in a number of different ways, including through parasitism, photosynthesis, and predation. Dinoflagellates are found in marine and freshwater, sediments, snow and ice, and even inside other organisms. They are an important food source for plankton, a critical component of coral reefs, and can periodically form potent harmful algal blooms. Heterocapsa circularisquama is an armored, free-living protist that can form extensive blooms and these blooms have resulted in severe financial losses to the aqua- culture industry. HcRNAV particles do not have an envelope and are approximately 30 nm in diameter. The linear RNA genome of this virus is only 4400 nt long. The two ORFs of the HcRNAV genome encode a protease, a replicase, and a single capsid protein. During the final stages of HcRNAV replication, the virus forms tightly packed crystalline arrays within the cell that result in the release of an estimated 20,000 viral particles during lysis.

Family Astroviridae

The family Astroviridae includes two genera. The capsids of astrovirids have nonenveloped, spherical particles that range in size from 28 to 30 nm and are covered in mushroom-like projections involved in virus attachment and entry. The genome of these viruses is a single linear molecule of RNA that is generally from 6400 to 7700 nt long and terminates in a poly(A) tail. Like the calicivirids, the astrovirids produce a subgenomic RNA that encodes the structural proteins. Astrovirid infections cause gastroenteritis in a variety of birds and mammals, including humans. In the marine environment, members of the Mamastrovirus genus infect Bottlenose dolphins, and California and Steller sea lions.

Family Caliciviridae

There are five genera of viruses within the family Caliciviridae. Viruses in this family produce nonenveloped, icosahedral virions 27–40 nm in diameter that has distinctive cup- shaped depressions. The genome of these viruses is linear, ranges from 7300 to 8300 nt long and has a poly(A) tail. Unlike some other taxa of group IV viruses, the calicivirids make a subgenomic RNA that encodes the structural genes during replication. Humans, pigs, rodents, and cats, among other vertebrates are all infected by calcivirids. Viruses in the genus Vesivirus have been found to infect diverse marine mammals, including sea lions, cetaceans, and walruses. Vesiviruses exhibit a broad cell tropism and a broad host range. For example, vesiviruses isolated from marine mammals are capable of infecting pigs and humans.

Family Nodaviridae

There are two genera in the family Nodaviridae. Viruses in this taxa are generally spherical in shape, (from 25 to 33 nm in diameter) and lack an envelope and discernible surface structure. The genome of nodavirids consists of two molecules of RNA (approximately 4500 nt in total length) that have 50 caps and lack poly(A) tails. Nodavirids are capable of infecting a variety of insect species including two species of Aedes mosquitos, in which they cause paralysis and death. Viruses in the genus Betanodavirus infect and cause disease in marine fish. Examples include flounder, grouper, and pufferfish, among others. In addition, several nodavirid-like viruses not yet approved by the ICTV infect prawns.

Family Togaviridae

This family comprises two genera, the alphaviruses and rubiviruses. While the Alphavirus genus has approximately 40 recognized members, the Rubella virus is the only member of the Rubivirus genus. The particles of these viruses are enveloped, spherical in shape and have an approximate diameter of 70 nm. The genome of a togavirid is a single strand of RNA that is from 7700 to 11,800 nt in length and has a 50 cap and a poly(A) tail. While most togavirids are transmitted via an arthropod intermediary, transmission of the alphaviruses that infect marine mammals and fishes do not appear to require a vector. Alphaviruses have been isolated that infect elephant seals, Atlantic salmon, and rainbow trout.

Group V: Negative-Sense ssRNA Viruses

Order Mononegavirales

The Mononegavirales is a large order of negative-sense single- stranded viruses that includes four families, two subfamilies, and 16 genera. Although the morphology of these viruses can vary widely (ranging from 80 to 805 nm in size), all virions are enveloped and most possess spike-like structures that are involved in attachment to and fusion with the host cell. The genomes of mononegavirads are linear and nonsegmented, range from 8900 to 19,100 kb in size, lack a 50 cap and a poly(A) tail, have a characteristic gene order (wherein from the 30 end, core protein genes are followed by envelope genes and then the polymerase), and encode an RNA-dependent RNA polymerase whose sequence is highly conserved. During replication, each cistron is transcribed utilizing a stop-start mechanism starting at a single promotor region located on the 30 ends. The Mononegavirales is a taxon of particular relevance to human health, because it includes virulent human pathogens such as the Ebola virus, rabies virus, and measles and mumps viruses, but also includes many representatives infecting marine organisms.

Family Paramyxoviridae

The paramyxovirids are viruses of vertebrates that are re- sponsible for notorious diseases in humans and animals such as measles, mumps, pneumonia, rinderpest, and Newcastle disease. The family is organized into two subfamilies comprised of seven genera. The virions of paramyxovirids are spherical in shape and approximately 150 nm in diameter. The genome of these viruses is a single molecule of negative- sense RNA that can be from 13,300 to 18,200 kb in length. All genomes of paramyxovirids encode a protein, creatively named "F," which causes the fusion of the virus and cell membranes at neutral pH. An interesting characteristic of the replication of paramyxovirids is the "rule of six." This rule alludes to the fact that the efficiency of RNA replication in these viruses is severely diminished if the number of nucleotides in the RNA template being replicated is not exactly a multiple of six. It is believed that the precise packaging requirements of the virus drive this surprising requirement. The Cetacean morbillivirus and the Phocine distemper virus are species in the genus Morbillivirus that infect marine mammals, while the Newcastle disease virus has been isolated from the common murre.

Family Rhabdoviridae

More than 200 different strains of rhabdovirids have been isolated from plants, invertebrates and vertebrates. The most distinctive feature of this family of viruses is that they all possess a bullet or rod-like morphology that can be from 100 to 430 nm in size. The genome of a rhabdovirid is a single molecule of negative-sense RNA that can be from 11,000 to 14,900 kb in length. Viruses in the genus Novirhabdovirus infect diverse fishes, including salmon, flounder, and eels. Two novirhabdoviruses of note, infectious hematopoietic necrosis virus and viral hemorrhagic septicemia virus, are responsible for the mortality of significant numbers of wild and farm-raised salmon each year. The rhabdovirid Dolphin rhabdovirus-like virus was isolated from a white-beaked whale in the North Sea.

Families Not Assigned to an Order

Family Orthomyxoviridae

The virions of orthomyxovirids are enveloped, spherical in shape, and range in diameter from 80 to 120 nm. They contain a negative-sense RNA genome composed of six to eight segments of different length that are in total from 10,000 to 14,700 kb in size. Because of their ubiquity and virulence, orthomyxovirids have been intensely studied since the early twentieth century. This research has resulted in an unprecedented understanding of the mechanisms of zoonosis and evolutionary history of these viruses. Analyses of genomic data suggest that wild aquatic birds may be the primary reservoir of the orthomyxovirids that now infect a wide range of vertebrates, including the human-infecting H1N1 strain responsible for the worldwide human pandemic of 1918. Of the five genera in the family, the genera Isavirus and Influenzavirus A have viruses that infect fishes and marine mammals.

Family Bunyaviridae

The family Bunyaviridae comprises five genera whose viruses infect vertebrates and plants primarily via an arthropod vector. Among other distinguishing features, bunyavirids have a genome comprised of three unique molecules of negative or ambisense ssRNA, that range in total size from 11,400 to 19,200 kb, that encodes four structural proteins. The particles of these viruses are pleomorphic and range in size from 80 to 120 nm. The terminal sequences of the three-genome segments are highly conserved among bunyavirids and play an important role during viral replication. Most bunyavirids are capable of replicating in both arthropods, where little cytopathogenicity occurs, and vertebrates, where they are cytolytic. Mourilyan virus is a bunya-like virus that infects two penaied prawns. Additional bunyavirids have been isolated that infect seabirds, including murres and puffins.

Group VI: Positive-Sense ssRNA Viruses that Replicate through a DNA Intermediate

Family Metaviridae

Metavirids blur the line between virus and retrotransposon in that only one of the species of virus in this family produces infectious extracellular particles (these virions are irregular in shape and range in size from 50 to 100 nm). Oddly, encapsidated particles are produced within the cytoplasm of the host cell and in most cases do not exit. Nevertheless, viruses in this family encode the hallmark genes (major capsid protein, reverse transcriptase, and integrase), share the same gene order, and have a similar genome length to retroviruses (5200–7600 kb). There are two known marine metavirids that infect the same species of pufferfish, the Takifugu rubripes Sushi virus (TruSusV) and Fugu rubripes Suzu virus (FruSuzV), classified in the genera Metavirus and Semotivirus, respectively.

Phenotypic and Genotypic Diversity of Marine Virioplankton

Detailed characterizations of cultivated viruses have been essential for developing a coherent viral taxonomy. This structure has brought order to the bewildering complexity of the viral world, but viral diversity as viewed through the lens of taxonomy has its limitations. The stringent requirements for establishing official taxonomic classifications mean that the process is slow and deliberate. The rate at which novel viruses can be identified far outstrips the rate at which they can be properly classified. Taxonomy also does not concern itself with relative abundance or biogeography, which is an issue of interest to ecologists. Microbial ecologists investigating viruses in the ocean have therefore been somewhat less concerned with taxonomic classifications than with practical measures of diversity. Some studies rely on cultivation (e.g., establishing host ranges), but many others have focused on quantifying abundance and diversity (phenotypic and genotypic) using cultivation-independent methods.

Host Range

The low-throughput of cultivation is a bottleneck in the characterization of viral communities, but the ability to propagate a virus in the lab allows one to collect ecologically relevant information that cannot be obtained any other way. One can determine, for example, in how many different types of cells the virus can replicate. Investigations of viruses with different host ranges can provide insight into the fitness costs that must necessarily be incurred to offset the benefits of being able to infect a broader range of cell types. Although viruses are often classified as having a broad or narrow host range, such a semiquantitative conclusion is necessarily conditional, since not every possible host can be screened. Despite that limitation, host range analysis has been very enlightening, because it can be exquisitely sensitive in discriminating between strains and it does not require sophisticated or expensive technology. To perform this analysis, each viral isolate is tested for its ability to lyse a suite of organisms. Since viruses tend to be more or less host-specific, the suite of organisms tested should primarily consist of different strains of the same species or different species in the same genus, but testing more distantly related organism allows one to identify viruses with particularly broad host range. The more organisms one tests, the greater one's chance of discriminating between closely related viruses. Use of this technique with marine bacteria has shown that the number of viruses infecting any given genus of marine bacterium is extraordinarily large.

Morphology

The subnanometer resolving power of the transmission electron microscope (TEM) allows one to discriminate fine structural details of individual virions. This power has been exploited to conduct qualitative surveys of marine viral diversity that hint at an extraordinary richness. Electron microscopy has been used for quantitative, comparative assessments of diversity in natural communities as well, but the resolving power cannot be fully exploited in this context for technical and economic reasons. Negative staining, which reveals the greatest detail of virions, is difficult to control and best suited to qualitative rather than quantitative assessments of diversity. Even with the greater uniformity provided by positive staining, it is difficult to confidently discriminate similar viruses in a mixed assemblage. A single virus may appear different depending on its orientation or as a result of damage or distortion during preparation. As a consequence of these limitations, quantitative surveys of morphological diversity have relied on classification into a handful of broad categories (e.g., <30, 30 to <60, 60 to <80, 80 to <100, and Z100 nm). The results of TEM surveys indicate virus-like particles in the ocean are usually numerically dominated by those with capsid diameters in the range of 30–60 nm, but range in size from around 20 to 750 nm. Viruses in the ocean thus span a similar size range as that observed among all classified viruses (18–500 nm). A significant, but variable, proportion of the viruses in seawater enumerated by TEM have tails, suggesting that they belong to the order Caudovirales and infect bacteria or archaea.

Genome Size

Genomes of viruses vary over orders of magnitude in length. Among marine viruses, for example they range from 3000 nt to 1,290,000 bp. Quantitative analysis of the genome length distribution in a sample can therefore provide some sense of diversity. Because conventional electrophoresis can only resolve DNA if it is less than about 20,000– 30,000 bp in length, pulsed-field gel electrophoresis (PFGE) has been used to analyze natural assemblages of viruses. The size range of virus genomes detected in seawater with this technique ranges from around 20,000 to 4500,000 bp. Under the right conditions, complex, fine-scale banding patterns can be seen, indicating that natural viral communities are quite diverse, but not all genome sizes are equally represented. In a wide variety of marine environments, the distribution of genome sizes appears to be multimodal, with peaks of genome copy numbers frequently occurring for sizes in the vicinity of 30, 55, and 60 kb. The peaks are broad, however, and there are bands detectable throughout the range. Bands 4100 kb are usually detected, meaning they contain a significant amount of DNA, but because there is more DNA per large genome, the copy number is low. The mean genome size of dsDNA viruses in seawater is around 50,000 bp, but the mode is closer to 30,000 bp. Because much of the DNA is clustered into several narrow size ranges, many genomes are not adequately resolved by PFGE. This method is therefore of limited value in quantifying viral diversity, especially richness, but remains a convenient means to track changes in com- munity composition.

Single-gene Surveys

Investigation of the diversity of microbial communities was revolutionized with the development of methods for cultivation-independent molecular surveys. Development of the polymerase chain reaction was transformative because it provided a means to amplify only specific genes of interest from a complex background of DNA. Analysis of the amplified material by sequencing or molecular fingerprinting allows one to determine how many variants of the targeted gene were present and thus evaluate genetic diversity. The small subunit ribosomal RNA gene has been the primary target for analyzing the diversity of cellular life, because of its many desirable properties including the fact that it is present in all cellular organisms. Viruses, however, do not code for ribosomal proteins. Viruses are so diverse, in fact, that there is no single gene that can be used as a universal viral target. Nevertheless, one can identify genes that are present and sufficiently conserved within subgroups of viruses that they can be used to survey diversity for a portion of the viral community. For example, structural genes of myovirids and phycodnavirids, DNA polymerase genes of podovirids and phycodnavirids, and the RNA polymerase gene of picornavirad have been used to assess the diversity of these various groups in the marine environment. Results from these surveys indicate that members of those viral groups are persistent and globally distributed, but within each group the specific sequence types detected and their relative abundance varies as a function of location, time, and depth. Comparing the viral gene sequences recovered by this cultivation-independent technique with homologous genes from viruses in culture has been very revealing. We now know that most of the viral diversity present in seawater is not represented in culture collections.

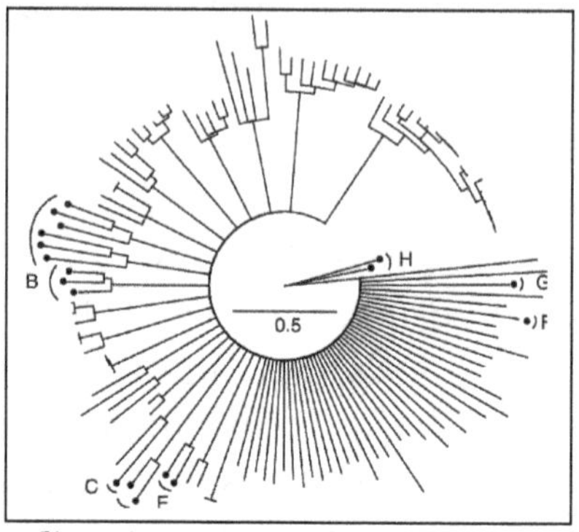

Comparison of the diversity among cultivated and uncultivated viruses.

The extensive diversity of viruses revealed by single-gene surveys and how it compares to the diversity among viruses that have been classified is illustrated using marine picorna- like viruses as an example. A phylogenetic tree was created using partial sequences of the RNA-dependent RNA polymerase gene for picornavirad strains in culture (branches marked with a black circle at the tip) or those identified from single-gene surveys of seawater samples. Groups of sequences identified by an arc and a letter contain representative members of established groups of picornavirads as follows: group A, Secoviridae; group B, Bacillarnavirus; group C, Labyrnaviridae; group D, Picornaviridae; group E, Dicistroviridae; group F, Marnaviridae; group G, Iflaviridae; and the outgroup H, Potyviridae.

Metagenomics

There are a number of limitations implicit in the targeted gene approach. First, one recovers information only about variants of a single gene (usually just a portion of the gene). Second, those genes come from only a subset of the viral community, and third, creating a meaningful target in the first place re- quires at least some prior knowledge about the viruses being sought. Metagenomic analysis, in contrast, allows for a relatively unbiased sampling of all of the viral genomes present in a sample, with no prior knowledge of their sequences. To analyze a viral metagenome, viruses are first harvested from a seawater sample, separated from cells and dissolved nucleic acids, then the viral nucleic acids are extracted. The extracted genomic material is then broken into random, small pieces for sequence analysis. For technical reasons, metagenomic analyses have so far tended to focus on either the DNA viruses or the RNA viruses present in a sample, but it is possible to analyze both at once.

After the analysis, one has a collection of sequence reads that derive from random fragments of the various viral genomes that were present in the original sample. Because one is sampling diversity across entire genomes, rather than a single gene, metagenomic analyses require much greater numbers of sequencing reads (thousands to millions) than are typical of single-gene surveys (tens to hundreds) to draw meaningful conclusions about diversity within a sample and how it compares to other samples. A first-order analysis one can perform is to compare each of the sequences that make up a metagenome to all of the sequences archived in various repositories to see if they are similar to genes that are already known. One can then classify each sequence according to its likely function and the type of organisms from which it might derive. In addition to analyzing each read independently, it is also informative to determine how many of the sequence reads derive from overlapping regions of the same genome, and can thus be reassembled into a single longer read. The higher the diversity in a sample, the lower is the probability that any two or more reads in the randomly selected pool of sequences that comprise the metagenome will overlap. By making a number of assumptions, one can estimate the diversity of the original sample from the assembly statistics for the metagenome. Although metagenomics has its limitations, it is providing many new insights into the genetic diversity of viruses in the environment.

DNA-Containing Viruses

In metagenome analyses of marine DNA-containing viruses, most (around 75%) of the viral genes do not resemble any gene with a known function or which derive from a known organism. Among those sequences that are recognizable, the largest proportion (90–95%) of the recognizable reads is most similar to genes of bacteriophages in the order Caudovirales. A much smaller proportion resembles ssDNA-containing phages and sequences from eukaryotic viruses, primarily those in the families Phycodnaviridae and Mimiviridae. Since these data are presumed to represent a random sampling, this suggests that the most common DNA-containing viruses in the ocean are viruses that infect bacteria, followed by viruses that infect protists. DNA-containing animal viruses are very rare in the virioplankton, but metagenomic analyses of samples taken directly from marine animals do reveal sequences resembling known animal viruses.

The percentage of viral DNA metagenomic sequences that can be reassembled into larger fragments is typically low. Statistical analysis of assembly data suggest that there are many thousands of distinct virus types in a cubic meter of coastal seawater and tens of thousands in a kg of coastal sediment. The shape of the rank-abundance curves for virus sequence types appears to be best described by a power-law function. Even the most abundant sequence type in these communities is estimated to make up only around 1% (water) or0.01% (sediment) of the total, which illustrates the exceptional diversity of marine viral communities.

RNA-Containing Viruses

The results from analyses of metagenomes prepared from viral RNA have been quite different than those prepared with viral DNA. A larger proportion of the reads (around 50%) are similar to known genes and almost all of them resemble eukaryote-infecting viruses. A small percentage of reads appear to derive from ribosomal RNA from cellular contamination, but none of the sequences resemble any of the known RNA bacteriophages. Most of the reads recognizable as viruses are most similar to sequences deriving from ssRNA-containing viruses in the order Picornavirales, but sequences resembling ssRNA viruses in other families outside of that order such as the Tombusviridae and dsRNA-containing reoviruses have also been detected.

Larger percentages of the sequences in RNA viral meta- genomes have been assembled than has been the case for DNA metagenomes. In fact, entire genomes have been readily assembled from random reads. Although RNA viruses are smaller than DNA genomes on average, the results also suggest that RNA viral diversity in the sea may be lower than that of the DNA viruses. However, there are no reliable estimates yet of just how diverse the RNA virus community might be and few environments have been sampled so far.

Marine Eukaryotic

Marine Fungi

It is important to consider that fungi evolved in the oceans in order to assess the diversity of marine fungi. While opinions differ as to when exactly fungi evolved from their ancestors, it is now generally accepted that the event took place at least a billion years before present. According to Heckman, this event probably took place around 1500 million years ago (Ma). The earliest plants, the chlorophycean algae appear to have emerged1100 Ma. It is also believed that plants conquered land much earlier than thought before, roughly around 1000 Ma, with the help of symbiotic fungi. Even if the exact dates of fungal origin are debated, fungi doubtlessly originated in the sea. This would mean that the lineages of early true marine fungi, which evolved in the sea prior to their conquest of land together with plants, probably continue to exist in the sea. As on A.D. 2000, about 450 species of obligate mycenean marine fungi have been described. Not many species have been described since then. Among the truly obligate marine fungi of the Kingdom Mycenae known at present, those belonging to the Ascomycotina and mitosporic fungi are the most common. Basidiomycetes and the earliest of true fungi, the chytrids, are few in number, the Zygomycotina being almost absent. When compared to nearly 70,000 species of fungi known so far on earth (out of 1.5 million estimated total number, a factor of ca.), the number of approximately 700 species of marine, mycenaean fungi is meagre. Even if the total number of obligately marine Mycenaean fungi is estimated to be 14,000 (20 times that of 700), this represents a disappointingly poor diversity of fungi in the sea compared to land, casting even serious doubts about their importance in the marine ecosystem. The straminipilan fungi belonging to Labyrinthulomycetes, which are truly marine, are more frequent in the oceans than the mycenean fungi, but even here the known number of species does not exceed about 50. The most 'primitive' of mycenean fungi, the chytridiomycetes are also extremely rare in the marine environment.

Two modern views of eukaryote classification

Cavalier-Smith[10]	Adl *et al.*
Kingdom Protozoa	Supergroup Amoebozoa Supergroup Rhizaria Supergroup Excavata
Kingdom Fungi (Kingdom Mycenae in this paper) Kingdom Animalia	Supergroup Opisthokonta First Rank Fungi (Mycenae in this paper) First Rank Mesomycetozoa First Rank Choanomonoda First Rank Metazoa
Kingdom Plantae Kingdom Chromista Subkingdom Cryptista Subkingdom Chromobiota Infrakingdom Haptista	Supergroup Archaeplastida Supergroup Chromalveolata Rank Cryptophyceae First Rank Haptophyta First Rank

Infrakingdom Heterokonta (containing the Hyphochytriales, Labyrinthulomycetes and Peronosporomycetes)	Stramenopiles (containing the fungal groups Hyphochytriales, Labyrinthulomycetes and Peronosporomycetes) First Rank Alveolata

Thraustochytrids are unicellular, produce plasma membrane extensions, the ectoplasmic net elements, and reproduce by means of zoospores. As is typical of the Kingdom Straminipila, these zoospores possess two unequal (heterokont) flagella, the longer anterior flagellum bearing tripartite hairs. They are also characterized by mitochondria with tubular cristae. Aplanochytrids may be unicellular or colonial and reproduce by spores that glide using ectoplasmic net elements, or by zoospores as above. Labyrinthulids are colonial and move within ectoplasmic net elements. Despite the fact that the first thraustochytrid was described by the great mycologist F.K. Sparrow nearly 70 years ago and that labyrinthulids have been known for a long time, only about 50 species of Labyrinthulomycetes are known so far. These fungi can be detected using the acriflavine direct detection (AfDD) method and cultured from natural samples using simple organic nutrient media. Several studies in recent years have demonstrated high abundance of Labyrinthulomycetes in the water column, using the AfDD method. We enumerated Labyrinthulomycetes in the water column of the Arabian Sea, using both the AfDD method, as well as culture methods. Cultured thraustochytrids amounted to only 0.06 to 3.65 % of the total thraustochytrids. In general, we have met with very little success in culturing thraustochytrids from oceanic waters.

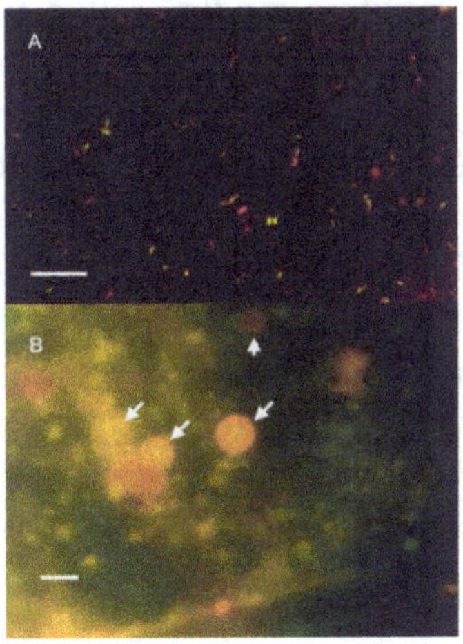

Difference between True Marine Fungi and Terrestrial Fungi

Lopez-Garcia examined the molecular diversity of eukaryotes within the range of 0.2 – 5 µm from the aphotic zone of 250 – 3000 m of the Atlantic polar front. Moon-vander Staay, likewise, studied picoplankton below the size fraction of 3 µm at a depth of 75 m in the equatorial Pacific. Their studies, using 18S rRDNA signatures,revealed the startling fact that fungal signatures were present in picoplankton samples.

The discovery of picoplanktonic, photoautotrophic cyanobacteria was discovered more than 50 years ago by Butcher in 1952 and was further confirmed by studies using microscopy and pigment composition. Further studies on the presence and importance of these phytoplankton of bacterial size, belonging to species of Synechococcus in 1979 and the discovery of species of Prochlorococcus in 1988 heralded research on picoplanktonic eukaryotes. Earlier studies on prokaryotic, autorophic picoplankton also revealed a complex assemblage of phototrophic eukaryotes of $< 3\,\mu m$. Some of the small eukaryotic photosynthesizers were subsequently described as members of a new order, Parmales. The smallest eukaryote ever found, Ostreococcus tauri, is a member of the Prasinophyceae and has a cell size of $0.8–1.1\,\mu m \times 0.5–0.7\,\mu m$. It contains a compact, tightly packed cellular organization comprising the nucleus, one mitochondrion and one chloroplast tightly. Such a cellular organization also appears in other photosynthetic picoeukaryotes such as the Bolidophyceae.

It was not till the two publications that heterotrophic eukaryotes of the picoplankton size were discovered. Lopez-Garcia found signatures of picoplanktonic true fungi at a depth of 3000 m in the Antarctic waters, picoplanktonic thraustochytrids at 2000 m and a new lineage of heterokonts, close to Oomycetes, at 250 m. Moon van- der Staay discovered molecular signatures of eukaryotic Oomycetes and an early heterotrophic divergence of stramenopiles in the Pacific waters. Significantly, the signatures observed in these waters revealed what appeared to be novel picoplanktonic organisms belonging not only to fungi, but also to the alveolates (ciliates and dinoflagellates) and heterokonts (stramenipilan organisms, including stramenipilan fungi). Among fungi, these authors discovered a novel group of sequences at the base of fungi and among the stramenipilan fungi, three phylotypes distantly related to thraustochytrids and labyrinthulids. Scanning electron micrographs of picoeukaryotes from 200 m deep Mediterranean waters showed cells in the range of 1 μm or less, one of these with heterokont flagellae, as in the stramenopiles.

It now appears possible that picoplanktoic single- celled fungi may be prevalent in the sea. If so, a fresh look at our conventional notions of how a fungus looks may need revision. It is also possible that marine single-celled mycenaean fungi have so far been placed wrongly as protozoans. Several members of choanozoans are a case in point. Choanozoans occupy a crucial position in understanding the origin of animals and fungi from a common ancestor. It has recently been shown that at least some of the members of choanozoans may actually lie at the base of fungal evolution. Thus, the single-celled nucleariids have been shown to be a sister clade to mycenean fungi. The single-celled fish parasites, the ichthyosporeans too are probably similarly related. The single-celled choanozoan protist, Corallochytrium limacisporum Raghukumar, actually possesses the fungal signatures ergosterol and the α-aminoreductase gene of the lysis synthesis pathway, thus displaying a relationship to fungi.

The mycelial form of fungi is suited to penetrate particulate organic matter and traverse space filled with air. Such a form is unsuitable for organisms living planktonically in the water column, a habitat that favours single-celled forms more. Therefore, it may well turn out that fungi at the base of the clade of the Kingdom Mycenae are single-celled and not mycelial and are typical inhabitants of the sea. A search for such organisms in the future may actually enhance our knowledge on the diversity of marine fungi.

Fungi in Extreme Marine Environments

Fungi have never been considered to be truly anaerobic in respiration. The only obligate fungal anaerobes known so far belong to the Neocallimasticales belonging to chytrids. No other such fungi have been discovered so far. Could anaerobic fungi be prevalent in the sea? The work of Dawson &

Pace using 18S rRDNA gene diversity in intertidal, black, anoxic sedments has revealed the extensive presence of eukaryotes in these habitats. These authors discovered numerous novel branches among the stramenopiles, including the fungal groups of oomycetes, hyphochytriomycetes and labyrinthulo- mycetes, as well as among protozoans. They further suggested that anaerobic eukaryotes may be the phylogenetically most diverse organisms in the sea. Likewise, Stoeck & Epstein studied the diversity of micro-eukaryotes in the suboxic waters and anoxic sediments of salt marshes. Most of the sequences represented deep, novel branches within green plants, cercozoans, and alveolates, as well as within mycenaen and stramenopilan fungi. Further work of Stoeck. on the permanently anoxic waters of the Cariaco Basin in the Caribbean Sea revealed organisms belonging to deep branches within the above groups and three novel lineages branching at the base of the eukaryotic evolutionary tree. These preceded, were contemporary with, or immediately followed the earliest eukaryotic branches. A significant number of these appeared to be new organisms. They suggested that these newly discovered protists might retain traits reminiscent of an early eukaryotic ancestor, at the base of the evolutionary trees. These exciting papers have a tremendous implication to marine fungal diversity.

Two of the exciting locations for investigations on anaerobic fungi in the Arabian Sea are the permanent oxygen minimum zone (OMZ) of the northern and central Arabian Sea and the seasonal OMZ of the west coast of India. It has been well known that the former is characterized by high densities of denitrifying bacteria. Jayakumar examined the diversity in the Arabian Sea OMZ of the prokaryotic 132 nirS gene sequences involved in the conversion of nitrite to nitric oxide. These authors found 12 major clusters, most of which did not show a high level of identity with other nirS sequences reported earlier. The dominant type in one of the surface samples was close to Pseudomonas aeruginosa. Highest diversity was found in samples with high nitrite. It is likely that fungi characterized by anaerobic denitrification also occur in such regions. Thraustochytrids are present in high numbers at these OMZ depths and their physiology is likely to provide us some surprise. Indeed, recent reports found several terrestrial fungi capable of converting nitrate to ammonia through denitrifying processes. Preliminary work carried out on such fungi in the coastal waters of Goa has already yielded several fungi, albeit belonging to known species, which are capable of denitrification.

Exotic Habitats of Marine Fungal Diversity

Several exotic marine ecosystems are yet to be studied intensively for fungal and other microbial diversity. Coral reefs, the deep sea and hydrothermal vents are some of the examples. Few obligate marine fungi have been reported from these habitats. For example, the Kohlmeyers described several marine fungi from dead corals. Surprisingly, not many have been added to this list since then, contrary to that of lignicolous fungi. Lignicolous marine fungi presently comprise about 450 species, new species having been discovered consistently since the seminal paper of Barghoorn & Linder in 1944.

Bacteria associated with invertebrates; especially in coral reef habitats is an exciting new area, of particular interest in the discovery of bioactive molecules. Although several fungi associated with sponges and producing interesting new molecules have been reported, all such species belong to terrestrial genera and few obligate marine fungi associated with such animals have been reported so far. This is a fertile area for future research. Similar is the case with deep-sea fungi. Terrestrial species of fungi adapted to deep-sea conditions probably play an important role in the ecology of deep-sea sediments. Fungal mycelia are easily detected in such sediments and fungi are probably prevalent in these habitats. Kohlmeyer & Kohlmeyer described obligate deep- sea fungi from decaying wood and bryozoan shells. Few have been described since then.

The evolutionary history of fungi strongly suggests that obligate marine fungi, unlike those reported so far from the sea are bound to exist. Biodiversity-rich ecosystems such as the coral reefs, the deep-sea and anoxic sediments are bound to harbour a high diversity of obligate marine fungi. Unfortunately few have been described so far. The paucity in our knowledge on the real diversity of obligate marine fungi in the above habitats brings us back to the questions that we asked in the beginning and we may conclude as follows.

1. We might have failed to recognize obligate marine fungi in exotic habitats because they look different from our conventional wisdom of fungi. Corallochytrium limacisporum Raghukumar, an inhabitant of coral reefs waters, the nucleariids and the ichthyosporeans discussed earlier are possible examples of atypical fungi in the marine environment.
2. Culture methods for marine fungi have been highly inadequate so far.
3. Molecular methods will provide us greater insights into marine fungal diversity.

Marine Protists

Protists are microscopic eukaryotic microbes that are ubiquitous, diverse, and major participants in oceanic food webs and in marine biogeochemical cycles. The study and characterization of protists has a long and distinguished tradition. Even with this history, the extraordinary species diversity and variety of interactions of protists in the sea are only now being fully appreciated. Figure below shows representative examples of marine protists, and of methods used to visualize these microbes.

Protists can be autotrophic or heterotrophic. The former, also called microscopic algae, contain chloroplasts, thrive by photosynthesis, and are at the base of all oceanic food webs, with the exception of deep-sea chemosynthetic ecosystems. There is a general trend for > 20-μm sized phytoplankton (microplankton), such as diatoms and dinoflagellates, to dominate episodically in coastal waters, while in the open ocean 2–20-μm sized cells (nanoplankton), such as coccolithophorids, sporadically form massive blooms that can be seen from space. However, during nonbloom seasons, even smaller cells (picoplankton— prokaryotes and eukaryotes less than a few micrometers in size) typically dominate phytoplankton biomass and production. Some genera of marine picoalgae can "bloom" to very high concentrations (> 105 cells ml-1), Ostreococcus, while others are ubiquitous, such as Micromonas, which is found from arctic to tropical waters.

In contrast to photosynthetic protists, heterotrophic protists have no permanent chloroplasts and rely on other organisms for nutrition. Most are phagotrophic— they ingest prey—usually other microbes. Some heterotrophic protists are parasites of phytoplankton or zooplankton, while still others form symbiotic relationships with autotrophic or heterotrophic bacteria. In addition, many species of protists have mixed trophic modes. These "mixotrophs" include flagellated phytoplankton that ingest bacteria or other protists, ciliates that "farm" chloroplasts from ingested algal prey, and mutualistic relationships between photosynthetic microorganisms and heterotrophic protists. Heterotrophic nutrition occurs among virtually all lineages of protists. Some protistan groups formerly classified as "algae" include species with strictly heterotrophic nutrition. For example, the dinoflagellates include many heterotrophic species, and it is possible that all dinoflagellates, including the photosynthetic species responsible for red tides, may have the capacity for phagotrophy.

Phagotrophic protists have long been known to be important in oceanic food webs as consumers of bacteria and phytoplankton, as regenerators of nutrients for further phytoplankton growth, and as a food resource for marine zooplankton. The grazing impact of phagotrophic protists has consequences for both ecosystem modeling and for the structure of microbial communities. The proportion of organic carbon produced by phytoplankton that flows through a multi-step microbial food web versus a shorter phytoplankton- mesozooplankton food chain is a critical factor that determines the capacity of marine ecosystems to sequester organic carbon and to efficiently produce fish biomass. Via selective grazing, both herbivorous and bacterivorous protists can alter the community composition of oceanic phytoplankton and bacterial assemblages. It is clear that these activities constitute major roles in oceanic ecosystems, and considerable information has been amassed on the abundances, distributions, and ecological roles of protists, yet there is still much to learn. At present, much of our knowledge is restricted to larger species that have readily identifiable morphological traits and those species that are amenable to laboratory culture.

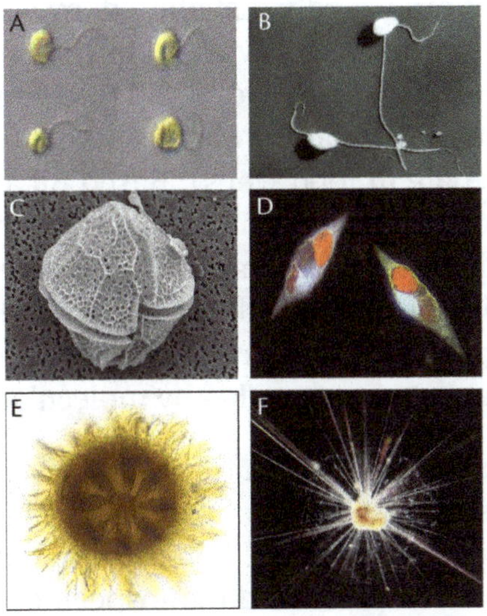

Examples of marine protists and of methods used to visualize eukaryotic microbes.

(A) Live cells of the 2-μm-diameter marine flagellated alga Pelagomonas calceolate, transmitted light microscopy. (B) Two cells of a 5-μm-long marine heterotrophic flagellate (Bodo sp.), scanning electron microscopy (SEM). (C) A 50-μm-diameter autotrophic thecate dinoflagellate (Lingulodinium polyedrum) that forms red tide blooms, SEM. (D) Two cells of a 20 x 60 μm marine heterotrophic dinoflagellate (Gyrodinium sp.) with food vacuoles full of ingested coccoid cyanobacteria, epifluorescence microscopy after preservation with formaldehyde and staining with the fluorochrome DAPI. (E) The 55-μm-diameter marine pelagic ciliate Strombilidium sp., inverted light microscopy after preservation with acid Lugol fixative. (F) Live foraminiferan Hastigerina pelagica with a fluid bubble capsule and a test approximately 300 μm across, darkfield light microscopy.

One fundamental question that has become a focus of research and debate in recent years revolves around the species concept (and species boundaries) for protists. Although protistan species have traditionally been defined based on their morphology, recent molecular analyses and physiological studies reveal that even well defined morphospecies may be composed of a mosaic of multiple genetic/physiological types. For instance, the ubiquitous diatom Skeletonema costatum, thought until recently to be a single species, is actually a composite of ten genetic types that are now considered distinct species. Similarly, physiological variability among Spumellalike heterotrophic flagellates indicates geographically distinct adaptations and emphasizes that morphology needs to be complemented by other approaches for distinguishing different species. The situation is even more complicated for the tiny picoplanktonic protists in the ocean. Molecular analyses during the last ten years have revealed many undescribed and uncultivated taxa among these morphologically similar forms, and whole lineages in the alveolate and stramenopile divisions have escaped detection until recently. The discovery of many cryptic lineages highlights the importance of combining molecular and ecological information with traditional morphological descriptions. This combination facilitates meaningful investigation of the distribution and activities of marine protists in situ.

These new approaches for defining protistan species have energized an ongoing debate regarding protistan species diversity and biogeography. The overarching question is whether there are relatively few species of protists that are broadly distributed, or whether similar protistan morphotypes are in fact distinct species or subspecies with more limited distribution. Why does species diversity matter? From an ecological perspective, it is important to know what taxonomic level (e.g., species or subspecies) yields information about associated differences in functional roles (i.e., how these organisms participate in food webs). Based on analysis of genetic variation, Katz found evidence for a "cosmopolitan" distribution (high gene flow and low diversity) for species of ciliates in coastal waters and evidence for "endemism" (high diversity and geographically restricted gene flow) among species of ciliates in isolated tide pools. Little is known at this juncture about the distribution, gene flow, diversity, ecological roles, or even the morphologies of the previously undescribed lineages of protists now being revealed by molecular analyses.

Beyond characterizing the diversity and distribution of protists in the ocean, major lines of research continue to elucidate the ecological roles of protists in marine ecosystems. Species-specific growth, grazing, and nutrient excretion rates of heterotrophic protists are derived mainly from laboratory experiments using isolated species fed monospecific prey under conditions that poorly mimic natural systems. However, such studies show that phagotrophic protists exhibit various types of species- specific behavior that may affect feeding behavior and growth, including chemosensory responses to prey and selective ingestion of prey types. In turn, prey cells may use structural or chemical defenses against protistan predation. Research on the underlying physiological/ biochemical basis of the feeding behavior of marine protists is yielding new awareness and understanding of their predator-prey interactions. The ability of herbivorous protists to discriminate between alternate prey types depending on size and "taste" undoubtedly affects protistan-grazing impact on bloom-forming phytoplankton, such as diatoms and harmful algal species. Studies of protistan herbivory on in situ phytoplankton communities indicate that food quality can play an important role in grazer selectivity and that protistan growth efficiencies vary greatly with prey type. Accurately describing the rates of activity of heterotrophic protists is extremely important to carbon-cycle models.

Unquestionably, many fascinating discoveries are ahead for protistan researchers at the physiological, organismal, and community levels. These include:

- Characterizing the overall breadth and meaning of protistan species diversity in the ocean;
- Understanding the ecology of minute organisms, such as the 2 μm heterotrophic flagellate Symbiomonas scintillans that itself harbors an endosymbiotic bacterium and the pico-alga Ostreococcus tauri, the smallest free-living eukaryote known, which can occur at high abundances and has unique genome features;
- Deciphering the ecological, molecular, and biochemical processes that might explain chloroplast acquisition by heterotrophic protists;
- Characterizing microbial interactions and processes sufficiently to provide a predictive understanding of community function and how microbial communities will respond in the face of environmental change.

Yet another challenge ahead is to understand the importance of processes such as parasitism and symbiosis in regulating oceanic plankton community structure and production; both are known to occur, but little is known about their prevalence and extent in marine systems.

Breakthroughs will depend on the application of a diverse array of approaches and methodologies. These will almost assuredly hold implications even beyond the important roles these organisms are already known to play in oceanic ecosystems. Marine protists likely retain some characteristics of the earliest eukaryotes that evolved on Earth. Interrogation of their genomes provides insights into how they thrive in the world's ocean and insights into fundamental biological processes such as the evolution of multicellularity, and thus will foster a better understanding of the evolution of life on our planet.

Seaweed

Seaweed is the common name for marine algae—a group of species from the Protista kingdom, meaning they are not plants at all, even though they may look like underwater plants, growing to more than 150 feet in length.

Algae are not plants, although they do use chlorophyll for photosynthesis, and they do have plant-like cell walls. However, seaweeds have no root system or internal vascular systems; nor do they have seeds or flowers.

Types of Marine Algae

Marine algae are divided into three groups:

- Brown Algae (Phaeophyta);
- Green Algae (Chlorophyta);
- Red Algae (Rhodophyta).

There is a fourth type of algae, the tuft-forming bluegreen algae (Cyanobacteria) that is sometimes considered to be seaweed.

Brown Algae: Phaeophyta

Brown algae is the largest type of seaweed. Brown algae is in the phylum Phaeophyta, which means "dusky plants." Brown algae is brown or yellow-brown in color and found in temperate or arctic waters. Brown algae typically have a root-like structure called a "holdfast" to anchor the algae to a surface.

One type of brown algae forms the giant kelp forests near the California coat, while another forms the floating kelp beds in the Sargasso sea. Many of the edible seawoods are kelps.

Examples of brown algae: kelp, rockweed (Fucus), Sargassum.

Red Algae: Rhodophyta

There are more than 6,000 species of red algae. Red algae has its often brilliant color due to the pigment phycoerythrin. This algae can live at greater depths than brown and green algae because it absorbs blue light. Coralline algae, a subgroup of red algae, is important in the formation of coral reefs.

Several types of red algae are used in food additives, and some are regular parts of Asian cuisine.

Example of red algae: Irish moss, coralline algae, dulse (Palmaria palmata).

Green Algae: Chlorophyta

There are more than 4,000 species of green algae. Green algae may be found in marine or freshwater habitats, and some even thrive in moist soils. These algae come in three forms: unicellular, colonial or multicellular.

Examples of green algae: sea lettuce (Ulva sp.), which is commonly found in tidal pools, and Codium sp., one species of which is commonly called "dead man's fingers."

Structure of Seaweeds

Thallus: the entire body of seaweed.

Lamina: a flattened structure that is resembles a leaf.

Sorus: a cluster of spores spore.

Air bladders: a hollow, gas-filled structure organ which helps the seaweed float, found on the blade). Other seaweeds (e.g. kelp) have floats which are located between the lamina and stipe.

Stipe: a stem-like structure, not all seaweeds have these.

Holdfast: a specialized structure on the base of seaweed, which acts as an "anchor" allowing it to attach to a surface (e.g. a rock).

Haptera: finger-like extensions of holdfast anchoring to benthic substrate.

Seaweeds play very important roles in many marine communities. They are a food source for many marine animals such as sea urchins and fishes, and form the base of some food webs. They also provide shelter and a home for numerous fishes, invertebrates, birds, and mammals.

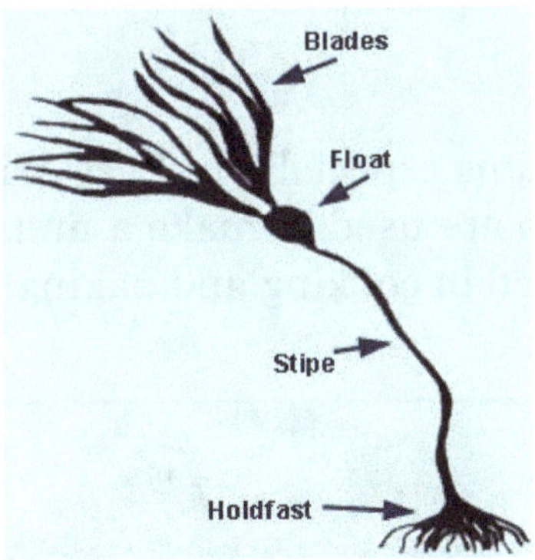

Structure of marine algae.

Seaweed Reproduction

Seaweed life and reproductive cycles can be quite complicated. Some seaweeds are perennial, living for many years, while are annuals. Annual seaweeds generally begin to grow in the spring, and continue throughout the summer. Some red seaweed has a life span of 6 to 10 years.

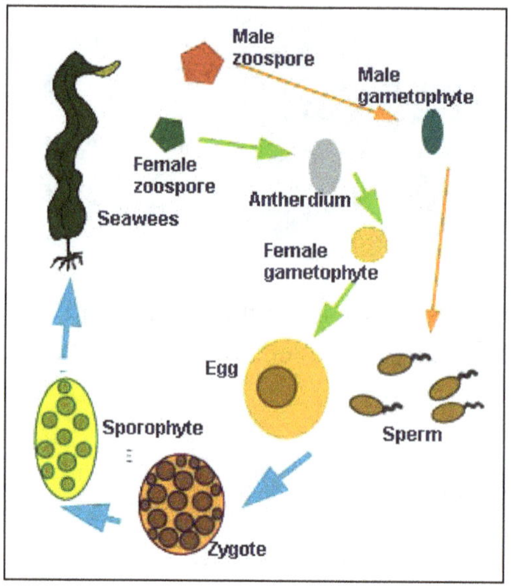

Seaweeds can reproduce sexually, by the joining of specialized male and female reproductive cells, called gametes. After they are released from the sporophyte, the spores settle and grow into male and female plants called gametophytes. The gametophytes produce gametes (sperm or eggs). The sperm and eggs are either retained within the gametophyte plant body, or released into the water. Eggs are fertilized when the sperm and egg fuse together, and a zygote is formed. Zygotes develop and grow into sporophytes, and the life cycle continues.

Seaweeds display a variety of different reproductive and life cycles and the description above is only a general example of one type, called alternation of generations. In a few species there is an alternating sexual and asexual reproductive process with every generation.

Seaweeds can also reproduce asexually through fragmentation or division. This occurs when parts of a plant break off and develop directly into new individuals. All offspring resulting from asexual reproduction are clones; they are genetically identical to each other and the parent seaweed.

Uses of Seaweed

Seaweeds are food source for humans especially in East Asia, it is most commonly associated with Japanese food. Seaweeds also are used to make a number of food additives such as alginates and carrageenan which is used in cooking and baking as a vegetarian alternative to gelatine.

Carrageean products.

Many seaweeds are used as medicine. Alginates are used in wound dressings and in the production of dental moulds and agar is used very widely in Microbiology to help grow bacterial cultures.

Agar plates.

Seaweeds are ingredients in toothpaste, cosmetics and paints and are used in industrial products such as paper coatings, adhesives, dyes, gels, explosives and many more.

Japanese food uses seaweeds extensively - Kombu a brown alga and Kim nori a red alga.

Nori.

Much of the oil and natural gas we use today formed from seaweeds which partially decomposed on the sea floor many millions of years ago.

Red Seaweed.

Prokaryotes

Marine Bacteria

Marine bacteria are single-celled organisms that can be shaped like little spheres, rods, or (less commonly) spirals. They are often very small, with cell diameters of just a few microns (about 1/100th the width of a human hair). They perform all kinds of chemical processes in the open ocean, including most of the steps in nitrogen cycling.

Cyanobacteria are a large group of photosynthetic bacteria, some of which can "fix" nitrogen, converting nitrogen gas into more biologically useful compounds. Cyanobacteria live in all kinds of environments, but are especially important in open-ocean ecosystems. They were formerly known as "blue-green algae" but are now recognized as a type of bacteria, not a type of algae (algae are eukaryotes).

Marine Bacteria that Fixes Nitrogen

Trichodesmium

Trichodesimium is a genus of colonial cyanobacteria that is one of the most important and well-studied nitrogen-fixing organisms found in open-ocean areas such as Station ALOHA. It is one of the few organisms involved in the oceanic nitrogen cycle that is visible to the naked eye. Trichodesmium is a colonial organism that forms hair-like strands, which sometimes aggregate into tiny "puffballs" up to a millimeter or two across. When winds are light, Trichodesmium colonies may clump together and float right on the sea surface, where they are known as "sea sawdust."

Trichodesmium uses an enzyme called "nitrogenase" to transform nitrogen gas into more biologically useful compounds (a process called "nitrogen fixation.") However, nitrogenase is inactivated in the presence of oxygen (which is produced by many photosynthetic organisms as a byproduct of photosynthesis). Because of this, most nitrogen-fixing microbes separate the processes of nitrogen fixation and photosynthesis either spatially (using different types of cells for each process), or temporally, by performing photosynthesis in the daytime and fixing nitrogen at night.

Trichodesmium, however, uses neither of these strategies. It is able to fix nitrogen during the daytime, but does not have specialized cells to perform the job. Researchers are very interested in figuring out how Trichodesmium is able to fix nitrogen in the daytime.

During the BioLINCS cruise, Trichodesmium is being studied using microscopy, nitrogen-uptake experiments, and nucleic acid (DNA and RNA) analyses. These analyses will be carried both out on the ship and later on shore. Among other things, researchers will be using genetic "probes" that look for a gene called nifH, which is one of several genes that help microbes produce the nitrogenase enzyme. This will help them determine the abundance and nitrogen-fixing ability of Trichodesmium. The Environmental Sample Processor (ESP) will also be using the nifH gene to monitor the presence of Trichodesmium in real time.

Heterocystus Cyanobacteria

Heterocystus cyanobacteria are multi-celled organisms that form microscopic filaments and perform nitrogen fixation in the open ocean. The most common genus of heterocystus cyanobacteria in open-ocean areas is Richelia, which is almost always found living inside of diatoms, a type of microscopic marine algae.

As part of this "symbiotic" living arrangement, the heterocystus cyanobacteria provide the diatoms with "fixed" nitrogen and other nutrients. It is not entirely clear what the bacteria get out of the arrangement.

Like all nitrogen fixing organisms, heterocystus cyanobacteria use an enzyme called "nitrogenase" to transform nitrogen gas into more useable forms of nitrogen. However, nitrogenase is inhibited in the oxygen-rich environment that exists inside most cells. As multi-cellular organisms, heterocystus cyanobacteria have evolved specialized cells called "heterocysts," which provide an environment more conducive to nitrogen fixation.

During the BioLINCS cruise, heterocystus cyanobacteria are being studied using DNA analyses that are carried out manually on board the ship and performed automatically within the Environmental Sample Processor (ESP). Among other things, researchers will be looking for the nifH gene in Richelia diatoms. This is one of several genes that allow cyanobacteria to produce the nitrogenase enzyme. Researchers will also be combining DNA analyses with measurements using influx flow cytometer (IFC) to determine whether heterocystus cyanobacteria are associated with specific types of marine algae.

Crocosphaera

Crocosphaera is a genus of singled-celled cyanobacteria that lives through photosynthesis and can "fix" nitrogen. It grows in many tropical ocean areas where the water is about 24° Celsius (75° Fahrenheit). At about two to four microns across, Crocosphaera is much smaller than Trichodesmium. Like all nitrogen-fixing organisms, Crocosphaera uses an enzyme called "nitrogenase" to transform nitrogen gas into more biologically useful compounds. However, nitrogenase doesn't work well in the presence of oxygen (a byproduct of photosynthesis). As a solution to this problem, Crocosphaera performs photosynthesis during the day and nitrogen fixation at night.

Uncultivated Cyanobacteria Group A (UCYN-A)

UCYN-A is a group of nitrogen-fixing cyanobacteria that were first discovered in the open ocean near Hawaii in 1998. Most of the time, they are not the most abundant nitrogen-fixing organism, but they can reproduce rapidly when conditions are right.

UCYN-A bacteria contain very few genes, compared with most other marine cyanobacteria. Although the entire UCYN-A genome has been sequenced, the organisms themselves have yet to be grown in the laboratory.

Surprisingly, sequencing of the UCYN-A genome revealed that that the bacteria lack genes that are necessary for "fixing" inorganic carbon and for key parts of the photosynthetic process. This allows UCYN-A to fix nitrogen during the daytime. However, because the bacteria cannot perform photosynthesis, they must have some other means of acquiring key organic compounds that are normally produced during photosynthesis. As with some heterocystus cyanobacteria, it is possible that UCYN-A form "partnerships" (symbioses) with other photosynthetic organisms to acquire these essential compounds.

During the BioLINCS cruise, UCYN-A are being studied using DNA and RNA analyses that are carried out manually on board the ship and performed automatically within the Environmental Sample Processor (ESP).

Alphaproteobacteria and Gammaproteobacteria

Proteobacteria are an extremely diverse group of bacteria. Some alphaproteobacteria and gammaproteobacteria, two subgroups within the proteobaceria, carry the nifH gene, and may be able to fix nitrogen. The proteobacteria also include some key players in the process of "nitrification," as described below.

During the BioLINCS cruise, the Environmental Sample Processor (ESP) will be looking for the nitrogenase (nifH) gene specific to gammaproteobacteria. This gene allows the bacteria to produce the enzyme nitrogenase, which they use to "fix" nitrogen. The ESP will also be looking for alphaproteobacteria using sandwich hybridization arrays.

Marine Bacteria Involved in Nitrification

Ammonium Oxidizing Bacteria (AOB)

The ammonium oxidizing bacteria (AOB) are bacteria that are all involved in a specific biochemical process (nitrification), but which may or may not be related to one another. They are involved in the first step of nitrification—the conversion of ammonium to nitrite (also known as "ammonium oxidation").

Most of the known ammonium oxidizing bacteria are betaproteobacteria and gammaproteobacteria. These bacteria are believed to be involved in this process because they carry a gene that codes for the ammonia monooxygenase enzyme (amoA).

Nitrite Oxidizing Bacteria

Several different genera of marine bacteria are involved in in the second step in the nitrification process—converting nitrite to nitrate (also known as "nitrite oxidation"). These include Nitrobacter,

Nitrospira, and Nitrospina Of these, Nitrospira is believed to be the most important in the open ocean.

Marine Bacteria that are Primary Producers but do not Fix Nitrogen

Prochlorococcus

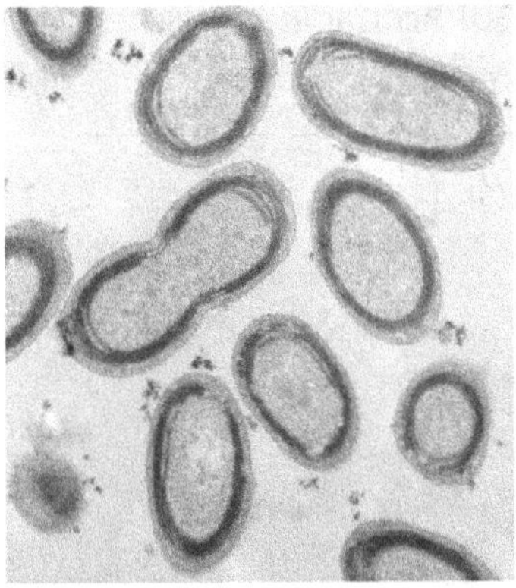

Prochlorococcus is a genus of cyanobacteria that is very common in open-ocean areas around the world. Although extremely tiny, with cells only 0.5 to 0.8 microns across, Prochlorococcus (along with another genus, Synechococcus) are so widespread and abundant that may produce a third of the oxygen in the Earth's atmosphere. Prochlorococcus are "obligate photoautotrophs," obtaining all of their energy through photosynthesis. However, they require nitrogen as a nutrient, and can use either nitrate or ammonium as a source of nitrogen. Thus, they are a key part of the "assimilation" process in oceanic nitrogen cycles.

During the BioLINCS cruise, researchers will be studying Prochlorococcus using the influx flow cytometer (IFC), and in some on-deck incubation experiments.

Synechococcus

Synechococcus is another common type of marine cyanobacteria. It is arguably the second most common group of photosynthetic marine bacteria, after Prochlorococcus. It has a similar shape as Prochlorococcus, but is typically a little bit larger, at 0.8 to 1.5 microns across.

Like Prochlorococcus, Synechococcus are "obligate phototrophs," which means that they can only obtain energy through photosynthesis, and require nitrate as a key nutrient. They can use nitrate or ammonium as sources of nitrogen, and are a key part of the "assimilation" process in oceanic nitrogen cycles.

During the BioLINCS cruise, researchers will be studying Synechococcus using the influx flow cytometer (IFC).

Marine Archaea

The Archaea are a curious phylogenetic domain (formerly kingdom) comprised of an odd assortment of cultured microbes that fall into three major groupings: extreme halophiles, methanogens, and extreme thermophiles and thermoacidophiles. Why such an odd assortment of salt-loving, or anaerobic, or heat-loving microbes should form such a coherent phylogenetic grouping is still not that well understood. The dogma until 1992 was that archaea inhabit mainly "extreme" environments, inhospitable to most other life forms. The existence of novel archaeal types was first hinted at during cultivation-independent ribosomal RNA surveys in open-ocean and coastal marine waters. Initial work used the polymerase chain reaction (PCR) to amplify ribosomal RNA genes from mixed microbial populations. In 1992, Jed Fuhrman of the University of Southern California first reported the existence of a new type of archaeal ribosomal RNA sequence from deepwater planktonic microbes in the Pacific Ocean.

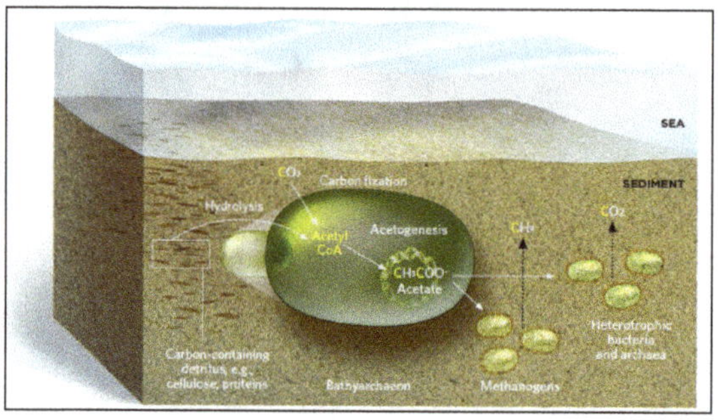

Archaea could be found in cold, aerobic habitats of coastal and open-ocean waters—and, to top it off, they were abundant. No cultivated, characterized archaea were known to grow at the combined salinity, temperature, and oxygen concentration found in temperate oceanic waters, shallow or deep. Following on the heels of these first oceanic sightings, archaeal groups began cropping up in many unexpected habitats. The next steps were to quantify the distribution and abundance of these unusual microbes and to begin to understand their biological properties and ecological significance.

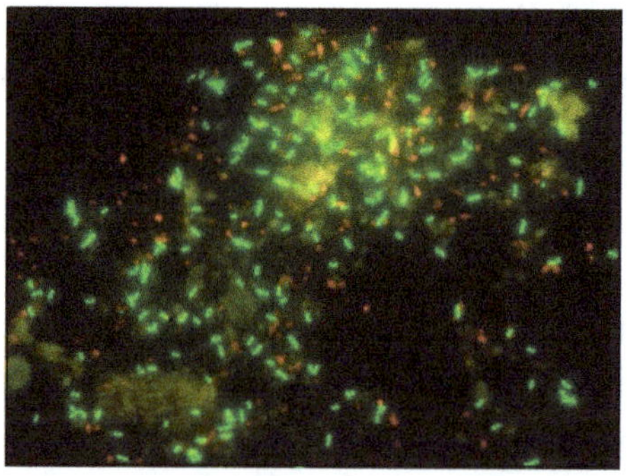

Marine Archaeal Abundance, Distribution, and Variability

Once it was clear that archaea were reasonably abundant players in microbial plankton, scientists initiated studies that employed radiolabeled, archaeaspecific oligonucleotide probes to quantify total, extractable archaeal rRNA in marine plankton. Surprisingly, planktonic Crenarchaea were found to contribute as much as 20% to the total microbial rRNA in late-winter Antarctic coastal waters at -1.8°C. Surveys in temperate waters off the coast of California also showed that the planktonic Crenarchaea tended to be most abundant in waters below the euphotic zone. This trend has generally held in a global sense in both coastal and open-ocean settings.

Ribosomal RNA-targeted fluorescence probes have been used to provide estimates of archaeal cell numbers in the water column. Off the California coast, for example, planktonic Crenarchaea represent > 20% of the total picoplankton cell counts from water depths of 80–3000 m. At the Hawaii Ocean Time-series (HOT) station ALOHA, Dave Karl and collaborators showed that Crenarchaea comprised as much as 30% of the total microbial counts in deep waters below the euphotic zone. In aggregate, these and other data suggest that pelagic Crenarchaea comprise a significant proportion of overall planktonic microbial biomass throughout the world's ocean.

Biology and Ecology of Planktonic Marine Crenarchaea

Creative application of biochemical, geochemical, and genomic techniques have provided considerable data on the planktonic Crenarchaea. These studies, combined with isolation in pure culture of a marine crenarchaeon (see below), now provide some specific clues regarding the biogeochemical and ecological importance of this abundant marine microbial group. Lipid analyses of cold marine sediments by Jaap Damste's group at the Netherlands Institute for Sea Research (NIOZ) revealed a first for this environment—high levels of tetraether lipids, that were previously found only in thermophilic Crenarchaea. Stuart Wakeham at the Skidaway Institute of Marine Science and my group at the University of California, Santa Barbara, then showed that marine plankton samples with high numbers of planktonic Crenarchaea also contained high levels of the same tetraether lipids. In addition, Preston showed that Cenarchaeum symbiosum, a crenarchaeal symbiont of marine sponges, contained the very same tetraether lipids. These could be used to infer the lipid's detailed chemical structure. Collectively, these data establish that the sources of abundant marine tetraether lipids in the plankton are indeed derived from planktonic Crenarchaea.

A real surprise came when Ann Pearson, during her graduate work with Tim Eglinton at Woods Hole Oceanographic Institution, provided radioisotopic data suggesting pelagic Crenarchaea may be chemoautotrophic. She purified large amounts of marine archaeal lipids from deep-sea surficial sediments derived from the deep-water planktonic Crenarchaea. Natural 14C isotope analyses of archaeal lipids suggested that deep-water archaea were not consuming much organic matter derived from surface primary productivity. Rather, deep-water Crenarchaea appeared to be using dissolved inorganic carbon as their main carbon source. Pearson's analyses are now supported by a variety of independent studies. For example, when Cornelia Wuchter, a graduate student at NIOZ, added 13C-labeled bicarbonate to a sample from the North Sea, then incubated the seawater in the dark, the heavy-isotope label from CO_2 was nearly exclusively incorporated into crenarchaeal lipids.

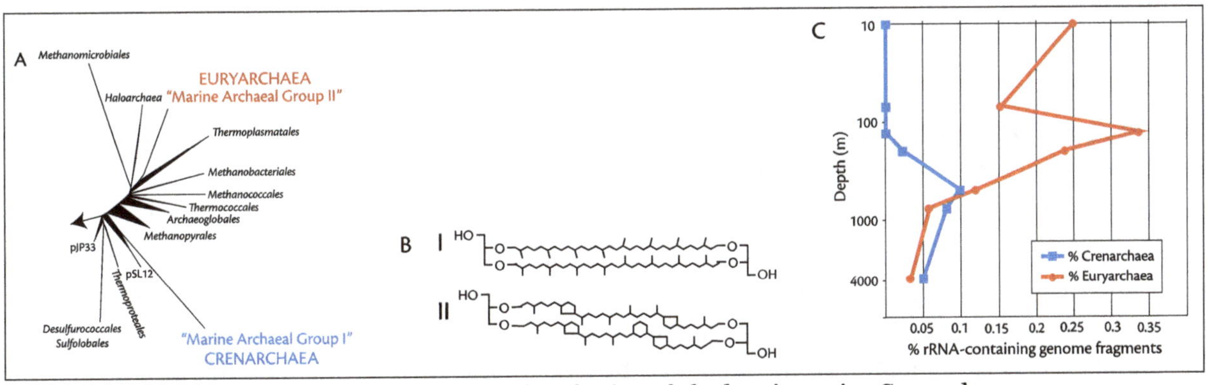

Phylogeny, lipids, and vertical distribution of planktonic marine Crenarchaea.

(A) A phylogenetic tree showing the general affiliations of two different planktonic archaeal groups. (B) General structures of abundant tetraether lipids found in Crenarchaea. (C) Distribution of planktonic Crenarchaea and Euryarchaea in the North Pacific Subtropical Gyre, as detected in large DNA-insert genomic libraries.

The observation of CO_2 -fixing Crenarchaea of course immediately led to the question: exactly what are these ubiquitous CO_2 -fixing marine Crenarchaea using as their energy source? Early clues came from the discovery of an ammonia monooxygenase gene in Crenarchaea that encodes a key enzyme used by nitrifiers in the first step of ammonia oxidation. This new archaeal ammonia monooxygenase gene was subsequently detected in many marine environments as well. Genomic analyses subsequently revealed many other genes associated with nitrification and CO_2 fixation in marine Crenarchaea. A definitive demonstration that marine Crenarchaea are indeed nitrifiers was achieved by Dave Stahl's group at the University of Washington, in collaboration with John Waterbury at the Woods Hole Oceanographic Institution. In cultures designed to isolate nitrifying bacteria (bacteria that oxidize ammonia obtain energy and utilize CO_2 as their carbon source), Stahl's group unexpectedly domain Bacteria. After some major microbe sleuthing, Stahl's group discovered that the new nitrifiers were, in fact, the same type of Crenarchaea that are so abundant in marine plankton. These crenarchaeal found ammonia-oxidizing cultures that did not appear to be typical bacterial nitrifiers—in fact, these nitrifying microbes did not even belong to the isolates grew exclusively on ammonia for energy, producing nitrite as the end product. The archaeal nitrifiers were also shown to use CO_2 as their carbon source, and they did not appear capable of growing on organic matter. These data, combined with earlier information on crenarchaeal distributions and abundance, indicate that planktonic Crenarchaea are indeed critical players in the ocean's nitrogen cycle, and they exert a large influence on oceanic nitrification in the sea.

New observations on the biology, ecology, and activity of planktonic Crenarchaea continue to accumulate. These include detailed quantitative analyses modeling in situ carbon sources of planktonic Crenarchaea, microautoradiography studies to track substrate assimilation into different cell types, and observations of Crenarchaea in anoxic zones of the Black Sea and the Arabian Sea. The potential interactions between Crenarchaea and other bacterioplankton have also recently been suggested by the similar distributions of bacterial nitrite oxidizers belonging to the genus Nitrospina and to Crenarchaea in the water column. Future studies promise to further elucidate the ecophysiological properties of the ubiquitous marine crenarchaeal nitrifiers and to better quantify and constrain nitrification and CO_2 fixation rates, as well as further characterize their ecological interactions.

Marine Microbial Diversity

Microbes were the only form of life for the first 2-3 billion years of planetary and biological evolution. Life most likely began in the oceans and marine microorganisms are the closest living descendants of the original forms of life. Early marine microorganisms also helped create the conditions under which subsequent life developed. More than two billion years ago, the generation of oxygen by photosynthetic marine microorganisms helped shape the chemical environment in which plants, animals, and all other life forms have evolved. Macroscopic life and planetary habitability completely depend upon the transformations mediated by complex microbial communities. These microscopic factories both aerobic and anaerobic are the essential catalysts for all of the chemical reactions within the biogeochemical cycles. Their unique metabolisms allow marine microbes to carry out many steps of the biogeochemical cycles that other organisms are unable to complete. The smooth functioning of these cycles is necessary for life to continue on earth.

Microbial life in the sea is extremely diverse, including members of all three domains of life (archaea, bacteria and eukarya) as well as viruses (most authorities do not consider viruses as living organisms). The members of these groups or taxa are distinct in terms of their morphology, physiology and phylogeny and fall into both prokaryotic and eukaryotic domains. Considering the adaptability of microorganisms to grow and survive under varied physico-chemical conditions and their contribution in maintaining the balance in ecosystems, it is pertinent to catalogue their diversity as it exists. The inability to visualize them with the naked eye precludes effective classification.

The diversity of microbial communities varies within habitats as much as between habitats. Marine microbial habitats can be classifi based on i) presence of other organisms (Symbiotic, Free living and Biofi ii) proximity to ocean surface or sediments (Euphotic: 0-150 m; Mesopelagic: 150 – 1000 m; Bathopelagic : >1000m; Benthos : sediments) and also based on iii) concentration of nutrients and required growth substances (Oligotrophic, Mesotrophic, Eutrophic). However, interfaces tend to be hotspots of diversity and biological activity. Marine microbial habitats at interfaces include the air-water, water-sediment, water ice, and host macroorganism-water interfaces. The sub-millimeter scale of physical and chemical variability in these habitats poses a serious challenge to studying interface habitats in detail. The fact that variations can even occur within a few millimetres, suggests that microbial diversity encompasses more than the documented evidence available. Hence, biogeography is gaining importance as a fi of study from microbial diversity point of interest. Due to the innately small size of the microorganisms, environmental complexity plays a major role in determining diversity. Spatial heterogeneity is likely to lead to the formation of many niches within a habitat. Recent tools like metagenomics aid in biogeography studies by providing information on nucleic acid sequence data, thereby directly identifying microorganisms. Therefore the phylogenetic information can be used to compare microbial diversity profi across habitats. Generally, diversity within a particular location and in a community is called alpha diversity. Beta diversity measures the community composition between two or more locations while gamma diversity applies to a region, across continents and biomes and is larger in size than that used for measuring alpha diversity.

Although microbial diversity is one of the difficult areas of marine biodiversity research, estimation of microbial diversity is required for understanding the biogeography, community assembly and ecological processes. The number of species has been a traditional measure of biodiversity in ecology and conservation, but the biodiversity of an area is much more than the 'species richness'. Diversity prediction can be made using statistical approaches that estimate species number from relatively small sample sizes. Hughes et al. (2001) noted that both rarefaction and richness estimators, which have been applied to microbial datasets, highlighted the utility of nonparametric estimators in predicting and comparing bacterial species number. Rarefaction and richness estimators rely on a species or operational taxonomic unit (OTU) definition. The limitations of this method are that OTUs are counted equivalently despite the fact that some may be highly divergent and phylogenetically unique, whereas others may be closely related and phylogenetically redundant. Recently, statistical analyses borrowed from population genetics and systematics have been employed and reviewed for use with microbial datasets to estimate species richness and phylogenetic diversity which do not rely on estimation of the frequency of different sequences. Reciprocal of Simpson's index (1/D), F-statistics (FST) and phylogenetic grouping of taxa (P tests) may be used as a measure of diversity, which has been widely used for ecological studies. These combined uses of species richness and diversity estimates provide information that enables deeper understanding of marine microbial diversity.

During the past two decades, molecular studies using the gene sequences that encode the small subunit rRNA (SSU rDNA) have revealed a wealth of new marine microorganisms that belong to the three domains of life viz., bacteria, archaea and the eukarya. The feeling among marine microbiologists is that of living through an age of discovery with no end in sight. However, most of this new diversity has not yet been described because pure cultures of the organisms behind the sequences are necessary to define a species. Recent studies of microbial diversity have produced spectacular discoveries of previously unknown microorganisms, many of which have major impacts on oceanic processes. Very large populations of picoplankton including diatoms, dinoflagellates, picoflagellates and cyanobacteria are the primary catalysts in carbon fixation, which orchestrate the cycling of nitrogen and form the base of the traditional marine food web. Heterotrophic SAR11 represents the dominant clade in communities of oceansurface bacterioplankton while nonphotosynthetic protists of unknown diversity control the size of picoplankton populations and regulate the supply of nutrients into the ocean's food webs. Communities of Bacteria, Archaea, and Protists account for greater than 90% of oceanic biomass and 98% of primary production. Archaea includes unusual microorganisms which grow under extreme environments and differs from bacteria due to lack of peptidoglycan. Both these domains collectively play a significant role in the marine environment.

Modern technologies (molecular techniques and automated fluorescence cell sorting) have demonstrated the great abundance and diversity of microbial life forms in the oceans, and DNA sequencing of environmental genomes (metagenomics) provides evidence of hitherto unrecognized physiological categories among the planktonic microbes. Since 99% of the microbial population is considered to be uncultivable, metagenomics assumes importance. Molecular techniques have identified SAR11 as a dominant clade in communities of ocean-surface bacterioplankton. Bacteria in the SAR11 clade (Pelagibacteraceae) make up roughly one in three cells at the ocean's surface. Overall, SAR11 bacteria are estimated to make up between a quarter and a half of all prokaryotic cells in the ocean. SAR11 bacteria are classified as alphaproteobacteria, and include the highly abundant marine species Candidatus Pelagibacter ubique.

Sequence information (eg. 16S rRNA sequences, genome sequences, metagenomes) as well as rRNA targeted probes (eg. Fluorescent In Situ Hybridization – FISH, which allows a visual inspection of phylogenetic groups of cells in a natural sample) have helped to discover many new groups of bacteria. Marine microbial genomics ranging from the study of the genomes of model organisms to the wealth of meta-omics approaches (e.g. metagenomics, metatranscriptomics and metaproteomics) has proven to be very successful to target the second basic ecological question on the role of microbes in the marine environment. Advanced technologies, developed in the recent past, promise to revolutionize the way that we characterize, identify, and study microbial communities. The most advanced tools that microbial ecologists can use for the study of microbial communities include innovative microbial ecological DNA microarrays such as PhyloChip and GeoChip that have been developed for investigating the composition and function of microbial communities. Single Cell Genomics approach, which can be used for obtaining genomes from uncultured phyla, enables the amplification and sequencing of DNA from single cells obtained directly from environmental samples and is promising to revolutionize microbiology.

Microbial Taxonomy

Currently, a polyphasic approach is used to define a microbial species using phenotypic and genotypic properties. Whenever a new taxon is proposed, it is essential that the organism be isolated in pure culture and its characteristic features be tested under standard conditions. All the strains within a species must show similar phenotypes. A designated type strain of a species constitutes the reference specimen for that species. If the 16S rDNA sequences of organisms are ≤ 98.5% identical, they are members of different species. Uncultured microbes cannot be assigned to a definite species since their phenotype is not known; however, they can be assigned a 'Candidatus' designation provided their 16S rRNA sequence subscribes to the principles of identity with known species. A concept applying to a taxon lower than that of the strain is the ecotype – those microorganisms that occupy an ecological niche and are adapted to the conditions of that niche.

Numerical Taxonomy

In numerical taxonomy many (50 to 200) biochemical, morphological, and cultural characteristics, as well as susceptibilities to antibiotics and inorganic compounds, are used to determine the degree of similarity between organisms. The coefficient of similarity or percentage of similarity between strains (where strain indicates a single isolate from a specimen) is then calculated. A dendrogram or a similarity matrix is constructed that joins individual strains into groups and places one group with other groups on the basis of their percentage of similarity. While 16S rDNA sequences have attracted attention in recent times as sole means of bringing out the uniqueness of a species; numerical taxonomy (based on phenotypic traits of a large number of species) compares favourably with that of genotypic data and, indeed, is in alignment with the modern taxonomy.

Polyphasic Taxonomy

In polyphasic approach of microbial taxonomy, phenotypic, genotypic and phylogenetic information are described as accurately as possible. The phenotypic information comes from the colony characteristics, cell type, cell wall-type, pigmentation patterns, proteins and other chemotaxonomic markers while genotypic features are derived from the nucleic acids (DNA/RNA). Phylogenetic information is obtained from studying sequence similarities of the 16S rRNA or 23S rRNA genes in case of bacteria and 18S rRNA in case of fungi. Many types of molecules are used for delineating and describing a taxon; some are mandatory (16S rRNA genes, phenotypes, chemotaxonomy) while others are optional (amino-acid sequencing of certain protein products, DNA-DNA hybridization), unless required for appropriate description. DNA barcoding approach is gaining popularity for assessing microbial diversity. Though only limited datasets (especially for eukaryotic microbes) are currently available, the scenario is improving due to faster and cheaper sequencing methods.

Factors that Impact Marine Microbial Diversity

Nearly every measurable physical, chemical, and biotic variable in the marine environment has been found to alter microbial diversity. The major variables affecting the marine microbial diversity include, turbulance, light, temperature, nutreints, salinity, pH, UV and solar influx, surfaces and interfaces, redox potencial, metals as well as presence of macroorganisms such as invertebates and macroalgae. For the most part, the extent to which each of these influences actually operates in the environment and the contexts in which they are important remain to be determined. Climate change, which will be felt by marine microbial communities as changes in ocean temperatures, will undoubtedly alter the diversity of communities in unforeseen ways. Climate change should be considered a major top-down controller of microbial communities. Pollution, including nitrogen inputs due to anthropogenic activities, also impact marine microbial diversity. Anthropogenic nitrogen inputs to the oceans now comprise about half the total nitrogen inputs to the oceans, a circumstance that has resulted in vast dead zones in coastal areas and an increased incidence of harmful algal blooms.

Marine Microbes and Bioactive Compounds

The marine environment is emerging as a 'gold mine' for novel bioactive compounds. Marine derived natural products present an enormous range of novel chemical structures and provide an interesting and challenging blueprint for creating new entities via synthetic chemistry. Marine invertebrates and plants, in particular, represent an environment rich in microorganisms that produce compounds with bioactive properties including antibacterial, antifungal, antiviral, anticancer, antifouling and antibiofilm activities. However, only 1% of these microorganisms can be isolated using traditional culture techniques, which has been a major bottleneck when mining the marine environment for novel bioactive molecules. Marine microorganisms have immense potential for providing services and products for human society which is not exploited to any significant extent. There is a world of microorganisms in the marine environment to discover, understand and put to good use.

Permissions

All chapters in this book are published with permission under the Creative Commons Attribution Share Alike License or equivalent. Every chapter published in this book has been scrutinized by our experts. Their significance has been extensively debated. The topics covered herein carry significant information for a comprehensive understanding. They may even be implemented as practical applications or may be referred to as a beginning point for further studies.

We would like to thank the editorial team for lending their expertise to make the book truly unique. They have played a crucial role in the development of this book. Without their invaluable contributions this book wouldn't have been possible. They have made vital efforts to compile up to date information on the varied aspects of this subject to make this book a valuable addition to the collection of many professionals and students.

This book was conceptualized with the vision of imparting up-to-date and integrated information in this field. To ensure the same, a matchless editorial board was set up. Every individual on the board went through rigorous rounds of assessment to prove their worth. After which they invested a large part of their time researching and compiling the most relevant data for our readers.

The editorial board has been involved in producing this book since its inception. They have spent rigorous hours researching and exploring the diverse topics which have resulted in the successful publishing of this book. They have passed on their knowledge of decades through this book. To expedite this challenging task, the publisher supported the team at every step. A small team of assistant editors was also appointed to further simplify the editing procedure and attain best results for the readers.

Apart from the editorial board, the designing team has also invested a significant amount of their time in understanding the subject and creating the most relevant covers. They scrutinized every image to scout for the most suitable representation of the subject and create an appropriate cover for the book.

The publishing team has been an ardent support to the editorial, designing and production team. Their endless efforts to recruit the best for this project, has resulted in the accomplishment of this book. They are a veteran in the field of academics and their pool of knowledge is as vast as their experience in printing. Their expertise and guidance has proved useful at every step. Their uncompromising quality standards have made this book an exceptional effort. Their encouragement from time to time has been an inspiration for everyone.

The publisher and the editorial board hope that this book will prove to be a valuable piece of knowledge for students, practitioners and scholars across the globe.

Index

www.ingramcontent.com/pod-product-compliance
Lightning Source LLC
LaVergne TN
LVHW080728210926
848485LV00003B/537